D1163993

PRAISE FOR *CYBER PRIVACY*

"In *Cyber Privacy*, April Falcon Doss has written the most sweeping, revealing, and understandable book about privacy and our digital lives. With up-to-the-minute analyses of everything from public health surveillance for COVID-19 to deep-fake videos flooding social media to how the NSA spies on terrorists overseas, *Cyber Privacy* is the comprehensive guide to our evolving notions of privacy in the digital age. A must-read if you want to understand how both businesses and governments know so much about you and how our society needs to adapt to preserve an individual's sense of identity."

— GLENN GERSTELL, **senior advisor, Center for Strategic and International Studies, and former general counsel, National Security Agency**

"We all have serious—but too often vague—concerns that every day computer usage poses a dire threat to our personal and financial well-being, as well the nation's security. In her new book *Cyber Privacy*, April Falcon Doss—the nation's leading expert on this subject—not only tells why that is so, but in a clear and engaging way arms us with strategies to protect ourselves, our loved ones, and the nation itself from life-threatening assaults on our privacy. This book is a must-read."

— MICHAEL GREENBERGER, **professor, University of Maryland Carey Law School and director of the Center for Health and Homeland Security, University of Maryland**

"A fantastically broad and uncompromisingly detailed tour through the world of data privacy. April Falcon Doss's perspective from the front lines takes us through the whirlwind of the past, present, and future of how our personal data gets used and abused by the people with access to it. From big tech companies, retailers, advertising companies, through to the police and intelligence agencies of the US and beyond, this is an absolutely critical read for anyone who wants to understand the complex, and often unintuitive, consequences of living in our increasingly data-driven world."

— MATT TAIT, **independent cybersecurity expert, formerly at GCHQ and Google Project Zero, and former senior cybersecurity fellow at the Robert Strauss Center for International Security and Law at the University of Texas at Austin**

"A brilliantly written tour de force on privacy in the 21st century. Combining decades of experience on all sides of the privacy debate, Doss combines incisive analysis of disruptive technologies, underlying economics, and increasingly complex legal overlays to deliver an essential primer on the fraught privacy landscape. Written in a straightforward style that takes no side, Doss delivers an enduring and useful framework for both individuals and societies seeking to align interests long thought to be beyond reconciliation."

—CHRIS INGLIS, deputy director, NSA, 2006–2014

"At a time when most internet users do not understand the complex concoction of algorithms, engagement, microtargeting, and personal data profiles that curate the information they see, April Falcon Doss uses her multi-sector experience to make privacy accessible to all. Anyone who cares about maintaining a grip on their personal information—or at least being informed about what's happening with it—should read this book."

—NINA JANKOWICZ, author of *How to Lose the Information War*

"Drawing on her experiences as an NSA lawyer, Senate lawyer, and private practitioner, April Falcon Doss has written a book that is unique in scope, well researched, clearly written, and a must-read for consumers, policy makers, privacy advocates, and concerned citizens. Without losing sight of the substantial benefits that are achieved through collecting and analyzing personal information on a massive scale, Doss exposes the unregulated practices of the large data collectors—including Apple, Amazon, Facebook, and Google—then examines the regulated practices of the Intelligence Community and the constraints—good and bad—on law enforcement activities. Her comprehensive discussion of both US and foreign laws drives the conclusion that our current approaches to privacy issues are often illogical, counterproductive, and even harmful. This book makes the case that we seriously need to re-examine what we are doing, and it provides useful guidance on where and how we can start to make meaningful changes that will benefit most everyone."

—DAVID C. SHONKA, former acting general counsel,
Federal Trade Commission, and privacy partner at Redgrave LLP

"April Falcon Doss has provided a vital contribution to our understanding of privacy and cybersecurity. *Cyber Privacy* provides laymen and experts alike with a rich understanding of the laws and technology that shape our ability to control who accesses our personal information and what they do with it. Drawing on her extraordinary experience in the public and private sectors, Doss takes an even-handed approach to examining our past, current, and future challenges, and presents complex issues in an accessible and entertaining format."

—JEFF KOSSEFF, **author,** *The Twenty-Six Words That Created the Internet*

"April Falcon Doss's book *Cyber Privacy* is a must-read for anyone interested in this fast-evolving topic. Whether you are a technology user, a compliance or privacy officer, or a practicing lawyer, this book will help in understanding the complex intersections of technology, the internet economy, the role of the state, and the uses of personally identifiable information and metadata. It's a particularly important topic right now, as regulations and laws are being developed and passed in jurisdictions all over the globe, and this very readable explanation of the major topics, players, and issues—current and future—is an essential guidebook."

—RICK LEDGETT, **former deputy director, NSA**

"April Falcon Doss thoughtfully, expertly and critically informs and navigates the reader across an amazing number of privacy invasion scenarios to an extent not seen in previous publications. She displays excellent insight and advanced analytical chops because of her many years of experience dealing with each of them. Novice and expert readers alike will profit from this important book. Bravo!"

—WILLIAM H. MURPHY, JR., **former judge and prominent civil rights attorney**

"April Falcon Doss has spent a career at the National Security Agency, Senate intelligence committee, and in private practice influencing the decisions that shape technology, cybersecurity, and data privacy. In this book, Doss turns twenty years of perspective and experience into a *Cyber Privacy* road map to guide those looking to understand how data came to rule our world and where we go from here."

—SUSAN HENNESSEY, **author of** *Unmaking the Presidency*

CYBER PRIVACY

CYBER PRIVACY

Who Has Your Data and Why You Should Care

APRIL FALCON DOSS

BenBella Books, Inc.

Dallas, TX

Cyber Privacy copyright © 2020 by April Falcon Doss

All rights reserved. No part of this book may be used or reproduced in any manner whatsoever without written permission of the publisher, except in the case of brief quotations embodied in critical articles or reviews.

BenBella Books, Inc.
10440 N. Central Expressway
Suite 800
Dallas, TX 75231
www.benbellabooks.com
Send feedback to feedback@benbellabooks.com
BenBella is a federally registered trademark.

Printed in the United States of America
10 9 8 7 6 5 4 3 2 1

Library of Congress Control Number: 2020014869
ISBN 978-1-948836-92-0 (trade cloth)
ISBN 978-1-950665-53-2 (ebook)

Copyediting by Michael Fedison
Proofreading by Lisa Story and James Fraleigh
Indexing by Word Co. Indexing Services, Inc.
Text design by Aaron Edmiston
Text composition by Katie Hollister

Cover design by Faceout Studio, Spencer Fuller
Cover photo © Shutterstock / Musa Studio (fin
  and honglouwawa (lock)
Printed by Lake Book Manufacturing

Distributed to the trade by Two Rivers Distribution, an Ingram brand
www.tworiversdistribution.com

Special discounts for bulk sales are available.
Please contact bulkorders@benbellabooks.com.

For Richard and Beverly Falcon,
who've been there since the beginning, and
whose love of reading sparked my own.

CONTENTS

INTRODUCTION
MAPPING THE PRIVACY
LANDSCAPE

When people and data-driven technologies intersect, the picture that emerges is a complex tangle of economic innovation, societal benefits, and damaging impacts on individual autonomy. At each intersection, data-thirsty apps and devices collide with the bundle of attributes that we commonly call "privacy." That single word serves as shorthand for a wide range of social values and individual prerogatives, such as the ability to control who knows what information about us and to limit intrusions into the solitude of our lives. Privacy certainly encompasses these things, but it also implies a great deal more. Privacy is intrinsic to individual dignity and our sense of personhood, to our ability to live as unique beings. Privacy allows us to test our ideas, to live without undue scrutiny. It lets us choose our relationships, overcome our pasts and direct our future, and change our minds and our behavior over time.

Data-driven technologies threaten to undermine all that. These are the technologies that lie at the heart of government surveillance, political manipulation, microtargeted commercial advertising, and the intrusive use of data by anyone seeking to exert power over others, including schools and employers, neighborhood associations, and even the intimate partners in our lives.

1

Understanding the risks that arise at these intersections is a challenge that grows more pressing every day. It's virtually impossible to live in modern society without leaving a digital footprint. Around 224 million Americans had a smartphone in 2018.[1] That's four out of every five American adults, up from just a third of Americans in 2011.[2] Some 64 percent of Americans have online accounts involving health, financial, or other sensitive data.[3] Nearly two-thirds of Americans have experienced some sort of data theft. Most Americans doubt the ability of the government or the private sector—and social media sites in particular—to protect their information. Roughly half of Americans think their personal data is less secure than it was five years ago.

But what can we do about it? Most Americans aren't taking basic steps to protect their information—and many aren't sure that they know what steps to take. Advertisers know what we're interested in and where we are; our social media profiles are hijacked to funnel political propaganda into our feeds; our employers have access to our health, wellness, and genetic information; and the very products and services—like cell phones and webmail—that we rely on for everyday life have become the basis for the most comprehensive corporate surveillance network the world has ever seen. Even when our data has been anonymized—when our names, addresses, and other identifying information have been stripped off—it's often possible for it to be "re-identified" through sophisticated computer analysis.

The urgency in understanding these issues is real. They're being covered by leading newspapers like the *New York Times* and *Washington Post*. They're entrenched in pop culture, from the ubiquitous hacker in modern movie scripts to recent documentaries like *The Great Hack*. The constant barrage of new developments in data privacy outpaces any one person's ability to keep up.

By taking a very human look across the intricate web of data to identify patterns and trends, however, we can take note of the major obstacles, and offer a road map that will help individuals navigate the privacy terrain. There are more privacy-protective routes that might be less efficient; more regulated routes that might involve extra costs and paying tolls; or routes that are congested with new kinds of data but that bring us within easy reach of all of the conveniences of modern life. Mapping the privacy landscape, as we'll do in this book, can help us choose our preferred solutions and figure out possible pathways to get there.

I've worked on issues relating to data, technology, and privacy for nearly twenty years. In 2003, when I started at the National Security Agency, "cyber-security" was a niche topic, the focus of IT professionals who still mostly referred to "information assurance" when describing their role in keeping computerized information safe. "Privacy" operated in a separate realm. Although a handful of privacy-related laws required organizations to keep some kinds of information free from prying eyes, many of those laws had been written with an eye toward analog systems. Legislators focused on the kinds of data that could be created and captured by regulated industries—like health care, financial services, or education—and laws often centered around information that wasn't of a fundamentally new type, but rather the kind that had previously been created on paper and now was stored in computers as a convenience or an afterthought. Much of the data collected about individuals was information that people could reasonably anticipate: we knew, for example, that doctors created notes from our visits, that workplaces kept personnel files, that the phone company whose switching system we relied on to communicate also, for its own business purposes, made a record of the numbers we called.

By 2016, when I left NSA, all that had changed. Smartphones, social media platforms, and ever-advancing digital technology had expanded the scope of information available about us, prompted the rise of whole new industries, and reshaped our interactions with information and with the people around us. The precision mapping applications on our phones track our movements in real time. Unbidden, they send us pop-up notifications suggesting the best route to a destination we haven't even searched where an algorithm predicts we will want to go. This is all based on the detailed records that have already been captured about our habits—where we go and when—over time. We unlock our phones and tag our friends on social media using sophisticated facial recognition software that identifies us in low lighting, from different angles, under varying conditions, and in wide-ranging contexts. We share video clips with our neighbors from the livestream on our front porch, which are also shared, sometimes without our knowledge, with the local police. We have digital assistants in our kitchens and our bedrooms, on our desks at work and in the classroom, all waiting to hear our next command—and recording our conversations in the meantime.

Moving from NSA into the private sector expanded the aperture of privacy and technology issues my work focused on. As the chair of the cybersecurity and privacy practice at a large law firm, I saw firsthand how changes in technology were creating as much uncertainty for businesses as for individuals: new types of data were being generated, and new techniques for manipulating that information were creating exciting new business opportunities; they also created new risks to individual privacy. The laws and policies that might have provided guidance on these issues simply weren't keeping up. Older privacy laws were often hard to implement or had been rendered obsolete in the digital age. New laws often imposed significant costs on businesses, but frequently without creating meaningful privacy benefits for individuals. And despite an abundance of laws, old and new, there were frequently gaps that left consumers and companies unsure precisely what their rights and obligations entailed.

In 2017, I joined the staff of the United States Senate Select Committee on Intelligence (SSCI) as the Senior Minority Counsel for the Russia Investigation. In that role, I gained an insider's view of the ways that data privacy intersected with pressing matters of political autonomy, propaganda, and foreign influence, and of the ways that shadowy third parties can use information about us to manipulate our opinions and divide our society. I grew increasingly concerned about the disconnect between the geopolitical magnitude of data's effects and the extent to which we're often willing to give up our data in exchange for nothing more than the convenience and enjoyment of free products and services, many of which are presented to us under conditions in which we don't know precisely where our data will go.

The topic of cyber privacy is evolving so rapidly that one of the greatest challenges in writing this book has been how to tackle the problem of recency: of making sure that the discussion remains fresh, current, and relevant, even while recognizing that no book can take the place of "breaking news." I have no doubt that, by the time this book is published, some new development in technology or the law will have emerged that isn't incorporated here. But achieving meaningful cultural awareness and comprehensive solutions to the ways that technology is eroding privacy won't come from chasing breaking news or from the breathless pursuit of the latest innovation (although I spend a great deal of time on Twitter, providing analysis on these trends as they emerge). Instead, this book aims to provide a thought-provoking overview of the ways that technology is challenging our notions of privacy, while also offering some thoughts about how to strike the right balance as a society: supporting important public safety and national security goals, enabling technology innovation to proceed, and allowing all of us to get our privacy back.

Informed digital citizenship shouldn't be a heavy lift; no one should have to devote their full time and energy to researching and understanding these issues unless they want to. It's my hope that this book will demystify the landscape of data-driven technologies so that you can discern which areas you care about the most. That journey starts with cataloging the major categories of personal data and unpacking the many subparts of what "privacy" means.

Are you worried about data collection in the private sector? From big platform providers like Facebook and Google, to the app you just downloaded on your phone, and the smart device you just installed in your home, this book explains the kinds of information that private companies are collecting about you and how they're sharing and selling it to others, often without your knowledge or deliberate consent. It explains the limits of US laws to govern, prevent, and regulate the actions of those companies and looks at the kinds of new laws being proposed, some of which might give consumers greater control over their data and others that appear to be privacy-protective but are unlikely to have any meaningful effect.

What about those who hold nongovernmental power over you? Employers? Schools? Intimate partners? What kinds of information can they collect about you? How is data being used to predict your employability, your anticipated success in school, or the likelihood that you'll cost an employer more on the company health insurance plan? Where are they getting this information from? You probably know that prospective colleges and employers look at your public social media posts. But what about tracking and analyzing the data on your fitness watch or reviewing the results of your genetic tests? Or using surveillance cameras to record what you do at work and at school? Many of the ways that personal data is collected and used might come as a surprise.

Is government surveillance your primary concern? This book describes many of the key tools and techniques used for government surveillance in the United States over the years, along with an explanation of the legal framework for those activities and how they've been implemented. It also examines the key risks and competing interests for the future and the ways in which lessons from intelligence oversight can be adapted to nongovernmental privacy challenges.

Although much of our focus will be grounded in the United States' legal framework, there are valuable lessons to be learned from the ways other nations approach privacy. Consequently, this book also examines privacy norms and data practices in key countries around the world. On the privacy-protective end of the continuum, the European model declares privacy to be a fundamental human right, but its regulations often increase costs without resulting in meaningful privacy benefits, and it overlooks some key areas of law—like European government surveillance—in favor of regional economic gains. On the other end of the international data protection spectrum, the world's most repressive, authoritarian regimes are leveraging new data-driven tools to monitor their citizens, crack down on free speech, and even to shape individual social standing. The international perspectives provide important guideposts for the United States in assessing the costs versus benefits of certain kinds of legislation and drawing lines between legitimate government use of personal data and the overreach that's clearly possible when personal information is misused in unfettered ways.

Finding the best path to navigate across a landscape of obstacles often means balancing economic growth and innovation, legitimate government purposes, and individual rights and liberties—to name just a few of the sometimes competing interests. In deciding the right balance to strike between those interests, reasonable minds can and do disagree. The goal of this book isn't to persuade you of any particular political viewpoint or of the importance of any specific approach to restricting—or leaving unregulated—the ways that data affects our daily lives. Naturally, there will be a number of areas in which my own opinions on those matters show through. But my goal is to give you a baseline understanding about key technologies and trends happening right now; a foundational knowledge about the current rules that do and don't govern those activities; and a framework for thinking about how we can shape laws and policies to reflect changing social norms. This book should help you set a privacy vision and present you with a road map for how we might get there.

SECTION I

What Kinds of Data Are We Talking About, and What Kind of Privacy Do We Mean?

CHAPTER 1

CATEGORIES OF DATA, AND HOW DATA IS COLLECTED

Throughout this book, we'll explore a number of definitions of "personal data" based on various laws, regulations, and policies. To get us started, however, it may be useful to think about what personal data means in the way that information security professionals often do: by *function*. Broadly speaking, personal data can be characterized as information about: 1) what we have, 2) what we do, 3) who we are, and 4) what we think, believe, and know.

This kind of personal information has been used to identify individuals, in some form or fashion, for decades (e.g., photo ID cards and computer passwords) or even millennia (e.g., portraits and watchwords in Greek and Roman antiquity).[1] However, changes in technology have created new ways to create, collect, store, and catalog data. Beginning with personal computers and widespread internet use, and accelerating with the advent of smartphones, the variety and amount of information in these four categories have expanded in ways that previous generations would have been hard-pressed to imagine. Thanks to current technology:

- personal data is created at ever-increasing volumes and rates;
- different data types often overlap or intersect and can be aggregated into multifaceted data sets;
- the combined data creates a vastly more detailed and nuanced picture of an individual than any single data input does alone; and
- advanced computer analytics use that information to draw sophisticated conclusions and make predictions.

Often, it isn't just what the data shows that threatens our privacy and autonomy; it's the inferences and conclusions drawn from that data, the assessments, predictions, and characterizations that get made about us from the digital breadcrumbs we leave behind. It's the power of aggregation, coupled with the ways that data-driven insights can be misused, that creates the greatest privacy risks. But before we get to that, let's first take a look at what each of these categories means, and identify some of the most prevalent, and impactful, examples among them.

It's the power of aggregation, coupled with the ways that data-driven insights can be misused, that creates the greatest privacy risks.

As a general matter, information about what we have is often the least intimate or sensitive, followed by information about what we do, who we are, and what we think, believe, and know. However, each of these categories incorporates a wide variety of information, informs and overlaps with the others, and can be used in countless contexts and multiple ways.

WHAT WE HAVE

In 2002, California became the first US state to enact a data breach law. The law was based on a consumer protection model and took a straightforward approach: it focused on information—like Social Security and credit card numbers—that, if compromised, could put a consumer at risk of financial fraud or identity theft, and it imposed a statutory obligation on entities that held that information to protect the data from unauthorized access. If there was unauthorized access—if the data was "breached"—they had to notify consumers, on the theory that people could then cancel credit cards or take other steps to protect their credit reports and financial accounts. The law was designed to address digital threats. First, it applied specifically to unauthorized access to computerized data, perhaps in recognition of the risk associated with storing large volumes of personal information and the reality that any computer connected to a network is also a computer that can be compromised or hacked. Second, the law provided a safe harbor for encrypted data: a breach that would otherwise trigger a notification obligation under the law became "no harm, no foul" if the personal data was encrypted and therefore effectively unavailable for exploitation by the unauthorized user.

Not long after California's law took effect, the concept behind this new data breach law began to spread to other states, which enacted similar laws of their own. Although the precise contours of each state's law vary, and many states—including California—have amended their data breach laws over time, the general premise has been recognized as an important mechanism for consumer protection. By 2018, all fifty states, along with Puerto Rico, the District of Columbia, and Guam, had enacted data breach notification laws.

These monetizable data elements remain at the heart of data breach laws in the United States today. Social Security numbers, driver's license numbers, bank account numbers, and payment card information are covered in some form or fashion by every data breach law in the United States, as well as by international data privacy laws and regulations. These data breach laws were among the first to explicitly tie individual privacy to computer security and to impose obligations not just on how the government handles information (restrictions that date back to the 1970s-era Privacy Act), but how anyone—including businesses and nonprofit organizations—protects information about who we are (identification numbers) or what we have (credit cards and bank accounts).

Over time, many other kinds of information that we "have" have become important as well. Where older versions of what we have tended to be more static—for example, bank account information and Social

Security and driver's license numbers—these new versions of what we have are both more dynamic, as we replace and upgrade devices on a regular basis, and more interactive, as the devices themselves generate new information about us on a near-constant basis. The quintessential example of something dynamic and interactive that most of us have is a smartphone or other computing device. All computing devices have some form of unique identifier that allows other networked devices to connect to them. These include the machine identifiers, or MAC addresses, that are unique to each device, including the internet-connected devices in our homes; the IMSI and IMEI numbers that uniquely identify mobile phones; the identifiers associated with the computer chips embedded in fitness wearables; even RFID chips implanted under our skin (more on this in a future chapter). All of these are potential sources of personal information, are vulnerable to unauthorized access, and, especially when combined with other data, can be used to generate information about our identities, activities, location, interests, and our inclinations—sometimes in ways we don't expect. The interaction between devices, their owners, and identifiers will be central to the discussion in later chapters about advertising technology and the data broker business. In the meantime, it's important to note that if we want to be anonymous in our online activities, it isn't enough merely to turn off cookies or log out of accounts (like webmail) that allow persistent tracking across sites and apps. We also need to consider the ways that who we are and what we do is tied to information about what we have. And, of course, from devices to Social Security numbers, there's still a lot of variety in what we have that can be used to identify us, be compiled with other data, and lead to compromises of other kinds of information.

WHAT WE DO

Many of our everyday actions are ones we carry out in plain sight: where we live, where we go to school or work, where we shop, and what items we buy. This behavioral information has been of interest—and of significant monetary value—to marketers for years.

Where a local butcher or greengrocer might once have known which customers preferred which kinds of items, the manager of a 30,000-square-foot grocery store that was open for sixteen hours per day, seven days per week, and owned by a regional or national chain was unlikely to know the names, much less the shopping habits or preferences, of any of their individual customers. The grocery store coupons printed in Sunday newspapers offered consumers savings that gave them an incentive to come into the store. But they provided the store with very little information in return. The store was rewarded with the purchases driven by that weekend's sale circular, but it didn't get the long-term benefit of substantially increasing its database of information about the individual purchasing habits of the customers who shopped there.

Creative executives identified a way to solve this marketing problem: enter the store loyalty program. With a free membership account, a customer could be guaranteed access to special members-only discounts at retailers of all kinds. The customers might welcome the opportunity to save money on their purchases, but stores also benefited by gaining valuable, monetizable information. Where traditional coupon programs only provided aggregate information about general response to particular discounts, loyalty programs augmented that aggregate information with detailed, per-person data on how specific shoppers responded to those offers. These programs and membership accounts also recorded every item customers purchased and tracked how their shopping habits changed over time. With this gold mine of information, stores no longer needed to make best guesses about what offers might be enticing, but rather they could provide individualized coupons—with certain coupons going only to certain customers—to increase the likelihood that specific customers would shop there in the future, spending more money with that retailer over time. By and large, consumers continue to view these programs as benign, and seldom ask what particular policies any given retailer might have with respect to the types of data they collect, how they use and protect it, how long they keep it, and who they sell it to or share it with.

Indeed, information that has always been visible has become so readily available and so transparent that it raises an important question: Now that so many types of personal data are available so cheaply and easily, has something fundamental in the nature of privacy changed? In other words, is this ready access to information merely a quantitative change? Or are there so many dimensions of "more," and are those dimensions so powerful, that it's now necessary to say that something *qualitative* has changed? Are we so subjected to continuous surveillance by corporate and government interests that we need to rethink privacy altogether? And do we need to create a new set of rules and expectations under which information that used to be non-private now becomes protected?

> *Are we so subjected to constant surveillance*
> *by corporate and government interests that*
> *we need to rethink privacy altogether?*

Although information like occupation, education, travel habits, reading preferences, and purchasing decisions have all been visible to others for quite some time, it's only in the past twenty years or so that the data has been so readily visible and accessible to nearly anyone who might want to view or acquire it. All of which brings us to the category of what we think, believe, and know.

WHO WE ARE

People have used biometric attributes to identify each other since the beginning of humankind. We recognize friends from a distance by the distinctive way they walk. We recognize loved ones even when their faces are concealed—bundled against the cold underneath a balaclava, or wearing a niqab out of religious or cultural modesty—based on nothing more than the distinctive shape and color of their eyes. We do this best with people we know; research on police lineups has shown how often errors are made when we try to match a likeness to someone we've only encountered once or twice, under stress.[2] Despite these limitations, our innate abilities work extremely well, and every day we rely on the expectation that we can positively identify people we know or have encountered before. Perhaps because we do this instinctively, we take for granted our ability to recognize people we already know and to create and store memory patterns for new people we meet, adding to our brains' personal biometric recognition database.

Beginning with the scientific enthusiasm of the Victorian era and accelerating through today, new technologies and analytical approaches have vastly increased the scale at which biometric recognition can take place, dwarfing the innate recognition capacity of any person acting alone. Using new techniques—photography, audio, video, or even radar—we're able to capture, record, catalog, and index specific biometric traits with increasing precision. Perhaps most important, when we need to compare or identify a specific biometric sample—a faceprint, fingerprint, voiceprint, or retinal scan—modern computer processing power allows us to quickly, cheaply, and easily compare a single specimen to the voluminous records stored in reference databases. When that information is coupled with the expanding network of sensors that constantly gathers information about us, we can even use biometric data to document or deduce a person's location and activities in real time.[3]

Most biometric information is so plainly visible that, historically, there have been few serious legal or ethical questions about whether a person's likeness, gait, or voice should be considered "private" in some way. Unless a person wears a full-face covering, their face is regularly displayed in public; unless they choose not to speak, their voice can be recognized by those who have heard it. When someone chooses to walk on the street or enter a shop or other public place, they know and generally accept that their actions could be seen. From early techniques like fingerprinting through facial recognition technology and genetic sequencing, biometric data raises some of the thorniest questions relating to cyber privacy.

FINGERPRINTS

Anecdotal evidence scattered through historical records suggests that societies have used fingerprints for identification on at least an intermittent basis for hundreds if not thousands of years. As far back as 2000 BCE, fingerprints were being recorded on clay tablets in ancient Babylon, and some accounts suggest that Babylonian criminals were being fingerprinted in 1700 BCE, during the era of King Hammurabi.[4] In China, fingerprints were used to sign clay seals in 250 BCE. The practice carried forward for millennia: fingerprints were used on government documents in fourteenth-century Persia.[5]

By the 1890s, a French anthropologist and police desk clerk developed what's believed to be the earliest systematic approach to biometrics. Alphonse Bertillon believed it was possible to use the precise measurements of a person's physical attributes—skull width, foot length, middle finger, and other bony structures—to uniquely identify them.[6] He further believed that, if at least fourteen of these separate measurements were taken, the chances of misidentification were 285 million to one. An ambitious approach, its effectiveness was frequently limited by human error: to be successful, this method required measurements to be taken with consistent technique and precision—not an easy task in nineteenth-century criminology.

Although Bertillon's approach didn't take hold, it laid the foundation for what would become one of the most widely used forms of biometrics: fingerprint analysis. Fingerprinting was already being used by British imperial officials in India who required colonial subjects, when signing a contract, to cover their palm and fingers in ink and impress them on the page. The practice achieved two goals: identifying the individual who signed the contract and intimidating subjects, a dual purpose that presaged the ways in which biometric information can be used today to reinforce the power imbalance between individuals and the state.

At the same time, fingerprints were beginning to capture the popular imagination and were being incorporated into criminal investigations. On the pop culture front, Mark Twain popularized the idea of using fingerprints to catch criminals in his 1883 memoir, *Life on the Mississippi,* and 1894 novel, *Pudd'nhead Wilson.* In law enforcement, what may have been the first criminal conviction based on fingerprint analysis came in 1892 in the sensational case of an Argentinian woman who was convicted of murdering her children, then slitting her own throat (she survived), and accusing someone else to cover up her crime.[7]

A tipping point had arrived, and fingerprint analysis took off. By 1903, New York State was using fingerprints to verify the identity of civil service applicants (and prevent cheating on the examination) as well as inmates in the state prison system.[8] The United States saw its first criminal conviction based on fingerprint evidence in 1912. By 1924, fingerprinting had become so widespread, for both criminal investigation and identification, that Congress passed legislation creating the Identification Division of the Federal Bureau of Investigation to hold the 880,000 fingerprint files already in the FBI's possession. By 1939, the FBI held more than ten million fingerprint records, and by 1946—thanks to fingerprinting of military personnel, defense-sector employees, expanded record-keeping relating to noncitizens, and continued criminal investigations nationwide—the FBI held files containing more than 100 million sets of fingerprints.[9]

As fingerprint technology has made it quicker, easier, and cheaper to capture, store, and compare fingerprints over time, the nature of fingerprint databases has also changed. Although early uses of fingerprints included validating transactions and confirming identities, the greatest impact of fingerprint collection was arguably in the investigation of crimes. Today, the FBI and the private sector possess vast stores of fingerprint, palmprint, and hand geometry information about millions of people who have never been investigated for, charged with, or convicted of a criminal offense. In some cases, such as crime scenes, people leave behind fingerprint evidence inadvertently. In other cases, people provide their fingerprints voluntarily and expect them to be used for wide-reaching background checks, for prospective jobs, professional licensing, or expedited airport security screening—or to be closely held, available only to the company they give them to, and only for limited purposes, such as using a fingerprint to sign into a gym or a doctor's office. And in still other cases, like using a fingerprint to unlock their smartphone, people are voluntarily using that piece of biometric data as they would set a combination on a safe: as a mechanism to protect a store of sensitive data, often on the assumption that the key can't be easily compromised by anyone—companies or governments—who might wish to access the information that the fingerprint protects.

The lessons of colonial India are instructive, however: when fingerprints and other forms of biometric information are collected, combined, and used today, whether by governments or the private sector, the risk of a power imbalance remains, a cautionary tale for anyone considering whether to voluntarily provide their fingerprint, whether it's to clock their time at work, unlock their cell phone, or access the speed-line for roller coaster rides at the local amusement park. Even when they are collected voluntarily for benign purposes, the unchangeable nature of fingerprints means that the information, if compromised, could end up being used for unforeseen—and perhaps exploitative—ends.

GAIT RECOGNITION

As far back as 350 BCE, Aristotle studied the ways that different creatures moved in his work *On the Gait of Animals*. In Victorian England, fictional detective Sherlock Holmes identified a murder suspect based in part on his analysis of the culprit's stride.[10] A person's gait is one of many attributes we naturally come to recognize in people we know—along with their hand gestures, the set of their shoulders, and the tilt of their head when they ask a question or listen to someone talk. Gait recognition has historically been an intuitive process. However, modern technology has made this tool more powerful and scalable than Aristotle or Sir Arthur Conan Doyle could have foreseen.

The turn of the twentieth century brought a leap in gait analysis due to the advent of motion picture recordings. Scientists found they could make simultaneous use of multiple cameras to record a person's movements from different angles. The resulting images revealed idiosyncrasies in the length of a person's step, the height of each footfall, and the relative angles of the ankle, knee, and hip joints. Scientists could even study the way a person's foot strikes the ground with each movement.[11] Taken together, these patterns created a distinctive profile of how a specific person walked.

It wasn't until the twenty-first century, however, that gait analysis became viable as a large-scale biometric technique. The availability of low-cost, high-volume data storage and cloud-based analytics made it possible to collect, store, process, analyze, and compare video files that previously had been too memory-intensive for automated manipulation. By 2008, researchers claimed that, based solely on imagery taken from satellites, they could identify specific individuals by their gait. Although the effectiveness of overhead photos and video was limited in certain situations, such as where they only showed a person's head and shoulders, a person's gait could even be analyzed, and recognized, if the video lacked good resolution on leg movement but showed the person's shadow.[12]

Gait recognition technology has become sufficiently advanced in recent years that it's coming into widespread use. Under a standard entitled "Biometric Information Management and Security for the Financial Services Industry," gait analysis can be used to identify individuals who are authorized to remotely access electronic systems in the financial services sector.[13] Recent studies showing that gait recognition analysis can accurately identify individuals as often as 99.3 percent of the time[14] have prompted interest in its use for airport screenings and have heightened interest in reviewing security camera footage in criminal investigations. It's already being used for public surveillance programs in China, where proponents of the technology have noted that "gait analysis can't be fooled by simply limping, walking with splayed feet or hunching over." It's easy to see why the technology would be appealing in a surveillance state: as one technology maker boasted, if a person can be recognized simply by how they move down a public street, "you don't need people's cooperation for us to recognize their identity."[15]

The advance of gait recognition technology underscores the impact that cameras have had on biometric information. In earlier times, a person might be seen walking down the street, but while their actions could be witnessed, they couldn't be recorded. With cameras came the ability to immortalize a moment, in some cases without the subject being aware; the privacy calculus began to change. The heightened discomfort that comes from knowing that someone else can quickly, cheaply, and surreptitiously make a permanent record of a person's presence or activities led, over time, to a number of developments that will be discussed later, including the definition of key principles of privacy law and lawsuits against social media behemoth Facebook for its photo-tagging suggestions. These factors have made the expanding use of camera-enabled surveillance one of the areas of greatest privacy debate, and facial recognition technology (FRT) lies at the center of the controversy. From laws designed to prevent Muslim women from wearing full-face coverings to prohibitions against masks in street protests in Hong Kong; from questions about the use of FRT in airports and police databases to the discriminatory impact of FRT in identifying persons of different ethnic backgrounds, the societal implications of FRT are expanding as quickly as the technology itself.

FACIAL RECOGNITION

It should be no surprise that, even now, at what we like to think of as an advanced stage of our evolutionary history, human babies have the ability to identify specific faces nearly from birth. By three months old, babies are more interested in looking at faces than at other kinds of images; they recognize and prefer their mother's face to other faces; they prefer faces that are the same gender as that of their primary caregiver; and if they are raised in a context where most of their caregivers are of a single ethnic group, the babies very early on demonstrate a preference for other faces within that ethnic group.[16] Taken together, all of these are hints from science that we're hardwired almost from birth to rely on facial recognition to recognize and differentiate between the individuals we know and trust and those we don't.

If facial perception is the innate human ability to recognize and make sense of features on human faces, then facial recognition technology is the set of automated tools that enable computers to make similar judgments. In some cases, computers can be more effective than humans at facial recognition. Not, largely, because computers are better at it—on the contrary, the human brain, which has evolved over eons to transform facial perception into a highly honed skill, is often more accurate in identifying faces than computers. But computers are able to store and access vast libraries of face-print examples and compare a particular face to those large data sets. If human beings were able to store in our synapses detailed information about the facial characteristics of thousands or millions of individuals, combining our accuracy with computers' volume would be a potent mix. However, since we can't, we're left to grapple with the significance of allowing computers to enhance—and sometimes err at—a task that our own unaided senses would be capable of carrying out on a much more limited scope.

The use of facial recognition technology and facial image records for identification goes back to at least the early days of photography—long before the days of computerized FRT capabilities. The photographic camera first came into use in 1839; within a few short years, the Pinkerton National Detective Agency claimed to be the first organization to take pictures of the people it apprehended.[17] Meanwhile, England began using "the angel copier"—prison photography—in 1852 in lieu of branding inmates, as a way to make it easier for police stations to share information with each other, and as a means to find and identify prisoners if they escaped.[18]

Semi-computerized facial recognition technology got its start in the 1960s, thanks to contracts from the US Department of Defense that supported pioneering work in the earliest phases of artificial intelligence by mathematicians William Bledsoe, Helen Chan, and Charles Bisson.[19] Bledsoe's team developed a system for comparing photographic images based on the distance between facial features such as eyes, ears, nose, and mouth.[20] This early FRT system required a combination of human and computer interaction: the human user entered data into a ten-inch-square electro-optical device that translated the inputs into ten thousand horizontal and vertical coordinates. With that detailed data entry, the tablet recorded the location and size of various facial features, along with the distances between them, and fed that information into a computerized database. When system users wanted to compare a new photograph with stored images to search for a potential match, the computerized system would search existing image files by looking for comparable distances between features, sizes of features, and similar attributes in order to identify the most likely matches.[21] Bledsoe's team noted at the time that this computerized system wasn't foolproof, and that any number of circumstances could confound its ability to make an accurate match. For example, if the same person was photographed from a different angle, that difference in perspective could make it appear as though the facial features were different distances apart, thus leading to the incorrect conclusion that the two photographs depicted different people. Nonetheless, this system, state-of-the-art for its time, expanded the potential for identifying individuals in ways that stretched far beyond what had previously been possible.

Facial recognition today has become far more sophisticated and advanced. It's being used at airports and concerts, in schools and stores, and to unlock our phones. Advocates of this technology argue that it will reduce shoplifting, financial fraud, and other kinds of crimes; that any privacy concerns can be managed; and that it will enhance users' experience when carrying out everyday activities.[22] Critics of FRT point out that it can be spoofed and hacked, leading to false results.[23] Facial recognition is also one of the forms of data collection that raises the greatest civil liberties concerns. Research has consistently shown that the accuracy of facial recognition algorithms varies widely across genders and racial and ethnic groups,[24] making disadvantaged groups especially vulnerable to misidentification by an FRT algorithm that misidentifies an innocent person as a criminal suspect. Because FRT can capture (albeit with errors) a person's race or ethnicity, many commentators fear that it could be used to offer different, and potentially discriminatory, prices to consumers based on their identity.[25] Free speech advocates fear that FRT can be used to thwart democratic protests—a fear that seemed well founded during the 2019 street protests in Hong Kong, when protesters adopted an ever-morphing series of adaptations to thwart the government surveillance cameras mounted on poles in Hong Kong's streets. In response, the art and fashion communities are continually devising inventive new ways, from jewelry to hairstyles to makeup, to foil the facial recognition technology used by public surveillance technology.[26] Even when used in purely commercial contexts, FRT raises unique questions about what permission platforms need in order to identify people, and who—the person or the platform—owns the mathematical model of an individual's face.

Cities around the United States have restricted the use of FRT in policing.[27] State legislatures have passed laws limiting the use of biometric data, including FRT. Courts have upheld lawsuits based on those statutes—leading Facebook to pay $550 million in 2020 to settle a lawsuit over its practice of using FRT to tag people in social media posts without their permission.[28] One activist and media commentator could have been speaking for many more when he noted that, "Facial recognition, or face surveillance, is a uniquely dangerous technology."[29] While this is undoubtedly the case, there are at least two more categories of "who we are"—medical records and genetic information—that also raise unique privacy risks.

MEDICAL INFORMATION AND RECORDS

Since 1996, US law has protected the privacy of certain kinds of personal data relating to medical and mental health diagnosis, treatment, conditions, insurance coverage, and related information under the Health Insurance Portability and Accountability Act (HIPAA). Although HIPAA had a transformative effect in standardizing the privacy of patient medical records, it was designed to meet very specific legislative goals relating to health insurance and the portability of medical information; consequently, the law operates within a very narrow context. As a result, the privacy protections required under HIPAA only apply to the specific categories of "protected health information" defined in the statute, and the law's privacy obligations only apply to "covered entities"—that is, to specific categories of businesses (such as hospitals, medical practices, and health insurers, and their "business associates," a category that includes companies like IT providers that handle information on their behalf).

In the quarter century since HIPAA was enacted, the volume, range, and types of health-related information that are being created and collected every day have expanded exponentially, and most of these new data types and new data stewards are not covered by HIPAA.

Perhaps the greatest growth of health-related information has come in the context of the increasingly popular fitness wearables, pioneered by Fitbit in the late 2000s with their clip-on activity tracker. Since then, more than a dozen other companies have been making devices that measure our steps and assess the quality of our sleep. These devices are often coupled with apps that let users track a range of health-related data. Some of it, like heart rate, is automatically collected by the wearable and sent to storage in the app's platform. Other information must be entered manually by the user: what the user ate, how much they weighed, the last time they had sex, and so on. Some of it is intimate information, like the data recorded by the 100 million women worldwide who use health-related apps to chart their periods and estimate their date of ovulation.[30] Other information may be less intimate but is equally revealing with respect to a person's physical and mental health: blood pressure and blood sugar readings, medical symptom encyclopedia search history, and interactive online chats with trained human listeners and AI chatbots.[31]

By and large, HIPAA doesn't apply to the companies that make wellness devices and apps or to the data they collect, whether that data is entered directly by users or collected automatically by the device. As a result, the information they collect—no matter how sensitive—isn't protected from being shared or sold to third parties by anything more than the privacy policies for the app or device. These policies are largely unenforceable by the users; they vary widely from one company to the next; and they often provide few details about how the data might be used, either on its own or in combination with other sources of information about the users.

GENETIC SEQUENCE

In the 1980s, the prospect of translating the roughly three billion base pairs in human DNA into a catalog of human genes seemed like a moon shot: highly ambitious, enormously costly, and with no guarantee it could be achieved. Yet in 1988 the National Institutes of Health announced the Human Genome Project, and the project was formally launched in 1990.[32] Research participation and financial support came in from around the world, with contributions from Australia, China, France, the United Kingdom, the United States, and others.[33] By 2003—two years ahead of schedule, thanks to an investment of nearly $3 billion—a rough draft of the genome was published. By 2006—$5 billion later—the final portions of the genome had been mapped, and the project was declared a success.

As with other areas of biometrics, technological changes were key to making a map of the human genome a reality. Improvements in genetic sequencing techniques were one key component, but advances in computing technology and computing power were equally vital to the project's success.[34]

By 2018, genetic testing had become so widely available that commercial companies like 23andMe, Ancestry.com, and others were offering genetic testing kits to consumers for as little as $129. They promised to provide insights into an individual's ethnic and regional heritage and connections to previously unknown relatives, and in some cases, to screen for genetic markers associated with certain medical conditions.

Genetic data is also surfacing in a number of law enforcement investigations, including ones in which the suspect didn't voluntarily provide genetic information, but family members—sometimes distant relatives—did. Perhaps the most well known of these instances was the 2018 arrest of the Golden State Killer.[35]

During the 1970s and 1980s, police were investigating a serial killer thought to have killed twelve women, raped at least forty-five, and burglarized more than 120 homes in communities across California. Despite those efforts, a suspect was never identified, and what became known as the Golden State

Killer case went cold for many years. A break in the investigation finally came forty years later, thanks to advances in analyzing DNA. For decades, police had retained a piece of evidence containing an uncontaminated DNA sample from the attacker responsible for one of the Golden State Killer's crimes. By 2016, the state of California had established a DNA database for persons arrested and convicted of serious crimes, and federal and state law enforcement authorities decided to reopen the Golden State Killer case, offering a $50,000 reward for information leading to his capture. Contra Costa County Detective Paul Holes used the decades-old DNA sample to create a DNA sequence map in a format that could be uploaded to the DNA matching database GEDmatch.[36] Based on the attacker's DNA sample, GEDmatch identified ten to twenty distant relatives, and a genealogist set about drawing the family trees that included individuals potentially matching the identity of the killer. From there, Holes and others began the legwork of narrowing down the field and zeroed in on Joseph James DeAngelo, then seventy-two years old, as a prime suspect. As part of the investigation, they collected DNA samples DeAngelo had left behind in places he probably didn't consider: from his car door, from a tissue left in his garbage, and from an item DeAngleo had been handling at work, provided by a coworker. Based on comparisons between DeAngelo's DNA and the crime scene samples, police arrested him on April 24, 2018. The rape and burglary charges could no longer be prosecuted because the statute of limitations had expired. However, on June 29, 2020, DeAngelo pled guilty to thirteen murders and was sentenced to eleven consecutive life sentences in prison, with no possibility of parole.[37]

There are a number of striking features about the way that DNA evidence figures in this arrest. First, the original sample was taken from evidence collected via a rape kit in the 1970s, which had been preserved for decades in law enforcement evidence storage. Second, narrowing the pool of suspects was made possible by DNA from distant relatives—DeAngelo himself hadn't provided his DNA to any of the commercial DNA testing services. But he had relatives who had, and their DNA markers were sufficient to draw the family tree that identified him as one of the potential suspects in the unsolved crimes. Third, the evidence that cemented his candidacy as a prime suspect was DNA evidence DeAngelo never intentionally provided; it was part of the constant trail of DNA that we all leave behind us in our everyday activities, on every object we touch and in every place we go. Under the law, there are almost no privacy or other protections against the collection and use of that kind of information. While most people would applaud the outcome in this case—the arrest of a serial killer, rapist, and burglar—it also points to the complexity of ethical issues relating to DNA evidence: both the promise of DNA information in solving crimes and the questions around collection and use of DNA information about individuals who don't know their data has been collected or who haven't consented to its use for particular purposes.

DNA analysis provides just one more example of the privacy intrusions that can result from biometric data, both because of the nature of the data itself and also because of the large-scale reference databases that must be maintained in order for there to be a meaningful set of data to search against. A key feature shared by all of these varieties of biometric information is that, unlike what we do or what we have, the biometric features that identify us can't be easily changed. Their immutable nature is what makes them so effective for identification and investigation, and is also why their collection and use create such privacy risks. Across the board, the advances in biometric identification and analysis raise a number of questions, including: 1) whether some kinds of biometric information (such as genetics) should be treated as being more sensitive than others (such as fingerprints); 2) how to remedy the harm to an individual when their biometric privacy has been violated; 3) how to counteract the impact of biometric conclusions that are biased by the ethnicity, gender, or other demographic features of an individual or group; and 4) how much privacy we are willing to sacrifice in order to achieve the cost savings, security, and convenience that proponents believe will be made possible through the use of these technologies.

WHAT WE THINK, BELIEVE, AND KNOW

This may be the most sensitive category of personal data. In many instances, concerns about other categories of information are really concerns that things we are, have, and do could provide insights into our interior lives: what we think, believe, and know. We know this intuitively: that intangible information about us can yield deep insights. For example, when critics of NSA's business records metadata collection program argued that there were privacy risks inherent in aggregating information about the existence of phone calls, the concern that many privacy advocates raised was not about the call detail records themselves, but about the risk that those records could be cross-referenced with other information that could identify the other parties to a call, and therefore serve as a basis for making assumptions, or reaching conclusions, about a person's thoughts or intentions. Put another way, the call records only revealed that phone number "A" had called, or been called by, numbers "B," "C," and "D," and the date, time, and duration of the calls. Critics worried that the government could theoretically combine those records with other information that identified the individuals or organizations associated with numbers B, C, and D, and then deduce (or assume) that person A was, for example, receiving care from a public health clinic known for treating drug addiction and HIV, that they were reaching out to a suicide hotline, that they belonged to the First Church (or Synagogue or Mosque) of Big City, or that they were active in Big Party politics.

Of course, all of that information isn't available from the fact that phone number A was used to dial phone number B, C, or D. But the concern shows how accustomed we've become to data aggregation. We assume that accurate and detailed directory information about the owners or users of phone numbers can be readily overlaid with the numbers themselves, and even superimposed on physical geography, allowing humans or algorithms to reach conclusions from the aggregated information. In some cases, that Hollywood-style idea of a Big Brother data society is greatly exaggerated. The ability to amass and combine those volumes of data requires significant money, sophisticated technology, and a permissive regulatory environment in which laws fail to constrain cross-data-set aggregation activities within limits that are consistent with American and other nations' values. Government organizations frequently have access to the needed money and technology, but in many cases are subject to significant regulation, oversight, and other constraints that prevent them from edging into authoritarianism.

In the private sector, however, fewer meaningful legal restrictions exist; as a result, companies that can hoard enough data, raise enough funds, and develop or acquire sufficiently robust computing platforms have at least the potential for deducing complex and sophisticated insights about what we think, believe, and know based on the digital detritus that our everyday activities leave behind.

Most of us have some intuitive sense of what "privacy" means in the physical world, but when it comes to the ways that our digital lives provide insights into what we think, believe, or know, our intuition often falters. For example, telephone metadata and cell phone location information are seen as highly sensitive. Yet, as individuals, we often don't consider the extent to which our internet searches and other online activities are being tracked, analyzed, and shared. The companies doing that analysis are at risk of leaping to unfounded assumptions, whose substance and impact on our lives could range from benign to humorous, disturbing, damaging, or dangerous. Indeed, marketers and behavioral profilers assume that our searches are intimately tied to the essence of who we are and how our opinions and purchasing habits can be shaped.

As it happens, the assumptions that these researchers draw from our online activity can be wildly off-base: browsing for beauty products leads advertisers to assume we're female; browsing retirement community information leads them to believe that we're elderly. When a profiler (or their algorithm) concludes that a particular person is a man between the ages of twenty-five and thirty-four, they're right as little as 25 percent of the time. For just a single characteristic, such as identifying a person's gender, the profiler is likely to get it right as little as 42 percent of the time. In both cases, the accuracy rate is worse than it would be from merely guessing.[38]

Accurate or not, most of these assumptions—and the results—are fairly benign. For example, if we use Google to search for weather and maps in another part of the country, Google's algorithms may conclude we plan to travel there and start serving us ads related to that location. If Google's algorithms are wrong, it's likely that the worst consequence we suffer is having "real estate" on our computer screen taken up by irrelevant ads. The more sensitive the search is, however, the greater the risk that erroneous assumptions about why we launched that search could have genuinely negative consequences for us.

Nowhere is this risk more evident than when, out of intellectual curiosity, we search for information on activities or conditions that are deemed dangerous or socially unacceptable. I might be searching for information on the viability of biological weapons or the steps to enriching weapons-grade nuclear material because those topics have been in the news and I want to understand more about how the technology works, how mature it is, and how easily a rogue actor could accomplish those goals. It's a purely academic interest on my part, but if I'm not cautious about how I search for that information, those searches will become part of the unique and voluminous aggregation of data about me. The same is true with searches for stigmatized medical conditions, disfavored social statuses, and any activity that could be seen as potentially dangerous or antisocial. When our search history becomes available to third parties, the risk increases that we will be judged not only for having made those searches but on *why* we searched at all.

In understanding the twin perils of how our search history can be misused and of relying on internet searches as reliable indicators of personal plans, intentions, or meaningful insights, it's important to remember how recent this phenomenon is—it wasn't until 2006 that "google" became formally recognized by the Oxford English Dictionary as a verb.[39] In an analog age, no one would have been able to track precisely what subjects an individual looked up in an encyclopedia. Now all of that information is catalogued each time we search online. And, as imprecise as the inferences may be, search strings often indicate *something* about our interests or our state of mind. Bad enough for data to intrude on the privacy of our activities and the state of our bodies; even worse for it to serve as some kind of mind-reading test.

So just how private should our search histories be? Reasonable minds differ, and the law on this isn't settled. But it's important not to overlook this category of information, whether it takes the form of the search terms we deliberately enter into a browser or the referrer links that show each website where we were on the internet before arriving, and which site we go to when we leave.

DEEP FAKES:
SYNTHETIC POLITICS AND PORN

The "deep fake" phenomenon is one of the most recent, and most rapidly evolving, data-driven technologies to arrive on the scene, going from a largely unknown digital experiment to widely available and sophisticated technology in just a few years. Deep fakes have been described as "synthetic audiovisual media,"[40] and "hyper-realistic digital falsification of images, video, and audio,"[41] and they show real people, often politicians or celebrities, appearing to do or say things that they've never actually done or said. The makers of deep fake videos rely on software that combines authentic digital information and enhances it with audio and video alterations that are made possible by artificial intelligence programs.

Deep fakes first emerged in a Reddit thread created on November 2, 2017.[42] Less than two years later, there were at least twenty websites and online community forums dedicated to the creation of deep fakes, and multiple apps enabled users to create their own.[43]

It's a staggering leap forward technologically. Where doctored videos in the analog age relied on deceptive editing—cutting and splicing audio and video recordings in ways that created false impressions—today's deep fakes allow video makers to create a wider range of deceptive content that never happened, such as the 2018 video that appeared to show President Barack Obama warning about the dangers of disinformation and using an expletive to refer to President Donald Trump.[44] That video, created by researchers hoping to warn of the dangers of deep fakes, went viral: it was viewed nearly five million times in less than six months.[45]

By late 2019, there were nearly 15,000 deep fake videos online. Ninety-six percent of them were pornographic, the top four websites alone garnering 134 million views.[46] Deep fakes today are generally created from digital representations of a person's likeness or voice, sometimes distorting the very information that they've posted to social media or provided to friends. They frequently rely on real video footage with someone else's face overlaid onto it, transforming porn footage, for example, by altering the actors' faces to look like celebrities like Taylor Swift,

Gal Gadot, or Daisy Ridley.[47] In other instances, the videos are a complete fiction, created from a pastiche of still images that have been enhanced by AI in order to create video of actions that never took place. A sampling of social media posts perfectly captures the chilling nature of what's possible. As one Reddit user said, "I want to make a porn video with my ex-girlfriend. I don't have any high-quality video with her, but I have lots of good photos." Another user on Discord posted that he made a "pretty good" video of a girl he went to high school with using a few hundred photos he had scraped from her Instagram and Facebook accounts.[48]

Deep fake porn videos are intended to harass, humiliate, and delegitimize the women who are targeted in them. There's a distinctly gendered dynamic in the content: fake porn videos overwhelmingly depict women, and often target specific women—female celebrities or politicians, ex-wives, or ex-girlfriends.[49] Other kinds of deep fake videos—purporting to show former President Obama giving an incendiary speech, or Nancy Pelosi appearing drunk—are intended for political misinformation and voter manipulation. For those non-porn topics, men and women are depicted at nearly equal rates.[50] In some cases, deep fakes aim for a perfect intersection of politics and misogyny, like the fake sex video alleging to be Hillary Clinton's that was pushed by Russian social media trolls.

As malicious as these videos can be, their potential for harm is mitigated somewhat by the shortcomings of the technology. It's often possible to tell that AI-generated video is fake; the eyes don't blink frequently enough to look natural, and the movement of lips and tongues isn't depicted in realistic detail.[51] As a result, many of today's deep fake videos have a look and feel that's a bit like foreign-language movies that have been poorly dubbed: the timbre of the voice might sound more like actor Jordan Peele than Barack Obama, and the movement of the lips and teeth might not quite line up with the audio track.

In the not-too-distant future, however, deep fakes will become more convincing and may be able to co-opt more than just our likeness and our voice. By the end of 2020, the technology may become so advanced that human viewers can no longer differentiate deep fake videos from real content—and our only solution will be to create other AI programs that can.[52]

In addition to producing nearly indetectable fake video, before long, software may be able to completely "spoof" our identities, to mimic our entire online persona in a highly convincing digital masquerade. Users will be able to upload digital images to a simulation program, creating an avatar that has an eerie resemblance to your face, your gait, your voice, and your mannerisms. The simulation software may pull data from other programs that track your internet browsing and location history, and that store your passwords and provide authentication for all of your financial accounts, email, calendar, and social media. With access to these programs and their data, malicious actors may be able to carry out the most convincing impersonations in human history.

As in other areas, the technology is moving faster than the law, leaving us with no comprehensive remedies for the digital duplication of our identities. Criminal laws address only certain prongs of this behavior: they prohibit hacking into people's computers and accounts, but most don't criminalize the misuse of digital data, like photos, that can be used to mimic someone's behavior, so long as the information was lawfully acquired. Civil lawsuits give a victim some recourse against a person who appropriates their likeness, but only if the victim can figure out who was behind it. Under take-down laws, an impersonation victim can demand that all copies of a faked video be removed from the internet, but only in some jurisdictions, and only for some kinds of content. By the time there's a criminal conviction or damages award, the harm—to reputation, to public perception, in fraudulent transactions, and in other ways—will have already been done.

While the complete digital spoofing of a person's entire identity isn't yet possible, the rapid evolution and spread of deep fake videos provide one more reason to be wary of the digital footprint we create, and to treat the gaps in the law as a matter of real urgency.

CHAPTER 2
A BUZZSAW OF BUZZWORDS

How Cloud Computing, Algorithms, and Analytics Are Impacting Data Today

M any of today's most sophisticated data uses haven't resulted from changes in the data itself; they've been made possible by advances in computer storage and processing. For example, complex medical diagnoses and predictions often rely on the same kinds of information that previously existed, handwritten, in patients' charts and researchers' notes. The difference is that today's computing technology allows all of the notes, charts, and research that had been recorded and held separately to be digitized and brought together onto cloud computing platforms, where they're combined with reams of related information and then crunched by algorithms looking for similarities. In some cases, these algorithms can make predictions in ways that, in the past, could only have been carried out by high-end, expensive, government-funded supercomputers. It's true that data-driven technologies of all kinds have vastly expanded the volume and types of information being created and collected. But many of today's thorniest questions about data privacy don't stem from the information itself but from the data analytics and predictive algorithms that cloud computing platforms make possible, available, and relatively cheap.

> *Many of today's thorniest questions about data privacy don't stem from the information itself but from the data analytics and predictive algorithms that cloud computing platforms make possible, available, and relatively cheap.*

WHY THE GROWTH OF COMPUTING POWER HAS BEEN A GAME CHANGER

Even in fast-moving areas of technology, some ideas have staying power. Gordon Moore put his finger on one that would become so influential, it became associated with his name.[1] Moore was one of the visionaries in the emerging field of computing: as cofounder of Fairchild Semiconductors and later the CEO of chip-maker Intel, Moore created and led one of the corporations that would reshape modern life.

In 1965, Moore published a now-famous paper in *Electronics* magazine about computer processing speed and integrated circuits.[2] In it, he predicted that the steady growth in processing power was likely to continue—and explained why it was poised to fundamentally transform the impact that technology had on society. According to Moore, since the advent of electronic circuits, the amount of information that could be processed by a circuit had roughly doubled every two years, and that pace was likely to continue for at least the next decade. Moore wrote:

> The future of integrated electronics is the future of electronics itself. The advantages of integration will bring about a proliferation of electronics, pushing this science into many new areas.
>
> Integrated circuits will lead to such wonders as home computers— or at least terminals connected to a central computer—automatic

controls for automobiles, and personal portable communications equipment. The electronic wristwatch needs only a display to be feasible today.

But the biggest potential lies in the production of large systems. In telephone communications, integrated circuits in digital filters will separate channels on multiplex equipment. Integrated circuits will also switch telephone circuits and perform data processing.

Computers will be more powerful, and will be organized in completely different ways. For example, memories built of integrated electronics may be distributed throughout the machine instead of being concentrated in a central unit. In addition, the improved reliability made possible by integrated circuits will allow the construction of larger processing units. Machines similar to those in existence today will be built at lower costs and with faster turn-around.[3]

In those few paragraphs, Moore predicted home computers, vehicle telematics, smart watches, the internet, and cloud computing.[4] All in 1965.

In fact, the growth rate that Moore predicted has accelerated, with processing power doubling approximately every eighteen months in recent decades. Computing power has increased a trillionfold between 1955 and 2015.[5] The average smartphone today has more memory and processing power than the computers that were used to send the first men to the moon. We haven't yet reached a natural limit posed by physics or mathematics—every time we think a limit is being reached, engineers and computer scientists have found new approaches to extend the realm of what's possible. When serial processing power was no longer enough, electrical engineers and computer scientists discovered the enhanced speed and capacity offered by parallel processing. When parallel processing was no longer enough, researchers developed the software and hardware frameworks that have made it possible to transition to the massively parallel capacity of cloud computing. Today, nations and corporations are in a race to develop quantum computing, a technology that's projected to be so powerful that it could dwarf current levels of miniaturization, speed, and power, making today's smartphones look no more sophisticated than the telegraph machines of a century ago.

If Moore's law continues to hold true, it's hard to anticipate what kinds of revolutions in data processing could become possible in the next five to fifteen years. What we do know is that, if the past serves as our guide, there's no end in sight.

HOW DATA HAS GROWN

By 2018, nearly 4 billion people around the world had access to the internet and were performing 5 billion searches every day.[6] Much of that data was being created on social media. Facebook, for example, reported 2 billion users around the world, including 1.5 billion users active on Facebook every single day, and posting 300 million photos daily.[7] Instagram—another social media platform owned by Facebook—boasted 600 million users, with half-a-billion of those users active every day, posting some 95 million new photos every twenty-four hours.[8]

The growth has been nonstop. In 2015, *Forbes* reported that more data had been created from 2013 through 2015 than in all prior recorded history.[9] In 2018, *Forbes* followed up that 90 percent of the data in the world had been generated in the previous two years—that is, between 2015 and 2017.[10] By April 2019, every day, internet users would post 500 million tweets, send 300 billion emails, upload 95 million photos on Instagram, and search the web 5 billion times.[11] By 2020, fitness wearables will have generated some 28 petabytes of data.[12] (It's often hard to visualize precisely how much a petabyte of data is, but roughly speaking, all of the information in US academic research libraries would only add up to 2 petabytes.[13]) It's useful to remember that Fitbit didn't roll out its first product until 2009, and the first smart watches didn't arrive on the market until 2012.[14] Although these devices are relatively new, their popularity has soared, with industry analysts estimating that some 245 million fitness wearables would be sold in 2019.[15]

WHAT COMPUTING POWER HAS MEANT FOR THE GROWTH OF DATA

As if in a feedback loop, the trillion-times growth of computer power has fueled the proliferation of data, and the growth of data has fueled the expansion of computing power. The two developments have been mutually reinforcing in speed and capacity.

> *The trillion-times growth of computer power has fueled the proliferation of data, and the growth of data has fueled the expansion of computing power. The two developments have been mutually reinforcing in speed and capacity.*

One of the key developments in computational capacity and data generation has been the rise of cloud computing. In roughly a decade, cloud computing went from a somewhat niche area of commercial and academic research to a household concept and name, with hundreds of millions of people around the world taking advantage of personal cloud computing accounts on services such as iCloud, Google Drive, Microsoft OneDrive, and other cloud storage platforms.

Like the internet itself, the concept for cloud computing originated in the Defense Advanced Research Projects Agency (DARPA), an agency of the US Department of Defense. The term first came into use in the 1960s and 1970s, with the concept of multiple users connecting to and sharing in the use of a single computer. By the 1980s and 1990s, computer scientists and engineers were regularly using the symbol of a cloud as a visual depiction for the series of computer connections between one user and another—in effect, a visual representation of an ellipsis, but for technical presentation rather than prose. The cloud was the symbol that went onto a drawing or slide to show that there were computer connections of some kind between here and there, but those connections weren't the central point of what was being captured on the slides.

Cloud computing surfaced as a transformative technology in the first decade of the new millennium, with two sets of papers and developments coming out of two multinational tech giants within a short span of time. In 2004, Google released a paper on the Map Reduce method of manipulating large-scale quantities of data, and in 2005, it released a paper on the Hadoop computing model. In 2006, Amazon Web Services rolled out its elastic compute cloud, and in 2008, Google released papers on its Google Files Service, Hadoop framework, and Map Reduce computational models. The release of these papers ushered in a new era of open-source and proprietary research on cloud computing and related technologies that spurred rapid developments in the field and laid the foundation for wholesale shifts in how computing technology would be used. Cloud architecture allowed multiple users to share storage space and processing speed and power on a network of multi-tenant servers, and computational frameworks like Map Reduce made it possible to collate and manipulate data across those scattered machines. Together, these new capabilities made it possible to bring unprecedented speed and power to indexing and analyzing record-setting quantities of data objects spread across multiple racks of equipment.

All of these technologies form the foundation for today's key privacy challenges: how data is collected and processed, what happens to the data that we generate as well as the information that others generate about us, and how that data is combined into an ever-growing stream of effects that impact our everyday lives.

If Gordon Moore's observation about processing power continues to hold true for another decade or more, we'll see this explosion of technology and data—and their intrusion into our lives—continue to accelerate at a pace even more rapid and head-spinning than what we've already experienced since the turn of the new millennium.

That prospect makes this an exceptionally good time to pause, reflect, and ask the questions: Who has my data? And why should I care?

Equally important are the questions that follow right behind those. What, if any, restrictions are there on how personal data can be used against me? And, with the rapid pace of new technology development, how can law and policy keep up?

CHAPTER 3
THE PRIVACY PRISM

A Single Term with Many Dimensions

P rivacy"—as we'll see throughout this book—doesn't have any single, simple, or agreed-upon definition in law, policy, or culture. This book looks at the notion of privacy primarily through two distinct lenses: 1) a Western democratic focus, shaped by English common law and the Western European traditions that have had the greatest influence on American culture, society, and expectations; and 2) through the perspective of technologies commonly used in the United States, the ways in which US law treats individual privacy, and the ways in which questions about data access are permeating American debate.

There are a number of reasons for this US-centric approach. First, many of the technologies that are most central to cyber privacy were developed in Silicon Valley. Second, perhaps because so many of those innovations have roots in America, the technologies have often been developed with an eye toward US law and cultural norms. Consequently, even where the technologies have been exported to a global user base, or where the technologies have been developed overseas, for many products and services whose business model depends on personal data, American laws and cultural norms exert a strong and sometimes defining sway over the mindset that goes into product development as well as data collection and handling practices. Despite its US-based perspective, however, this book also considers European data protection laws and their influence on American-based global technology companies, as well as the data practices of countries like China and Russia, which strictly limit the types and amounts of information that their citizens can access, and which use personal data as an intentional means to coerce and constrain individual speech, activities, and thought.

Bearing in mind the inevitable cultural differences that give rise to differing concepts of privacy around the world, any attempt to understand how technology is reshaping our notions of personal privacy has to start with the question: What does "privacy" mean?

FOUNDATIONS OF PRIVACY IN THE UNITED STATES

The United States legal framework recognizes several key sources of law: court decisions that have descended from English and American common law; the principles identified explicitly and implicitly in the US Constitution and Bill of Rights; and federal and state statutes. The US Constitution doesn't contain any language explicitly recognizing a right to privacy—in fact, the word "privacy" doesn't appear anywhere in the Constitution or Bill of Rights.[1] However, English common law provided important context that allowed American courts to define a right to privacy in the nineteenth century.

Under early American law, the rights that today we consider to be part of privacy were closely tied to property ownership: the things people did in their homes were private because homeowners had a right to keep other people—and their nosy inquiries—out of the homes they owned. Correspondence was private because property ownership principles meant that only the sender and recipient had the right to break an envelope's seal. American colonists fought a war, among other reasons, to ensure that citizens of the newly formed country could refuse to turn over the contents of their letters to the government and refuse to allow police to search their homes. These protections are at the heart of the Fourth Amendment to the Constitution, which declares that the government may not violate the right of the people "to be secure in their persons, houses, papers, and effects," and which requires any government search of those things to be reasonable, often based on certain indicia such as warrants. Although the Fourth Amendment applies only to government actions, it was also the case under common law that, if a private citizen invaded another person's home or read their mail, the rightful owner of those things could sue the intruder—not based on a claim of invasion of "privacy" as we understand that notion today, but on a claim of trespass.

Throughout most of the eighteenth and nineteenth centuries, the right to privacy remained largely tied to conceptions about property ownership. By the late 1800s, however, the realities of urban, industrialized life and innovations in technology were intruding into people's lives in ways that prompted courts to react and recognize a new right to privacy that didn't depend on owning property. Although the Fourth Amendment protected people from government intrusion into protected spheres like homes and property, it couldn't address the increasingly vexing question of whether the law should recognize rights to recourse if private actors interfered with a sphere of life that didn't yet have a legally protected name but which seemed to involve similar interests relating to the ability to shield personal information and actions from prying eyes, and the ability to keep strangers at bay.

Two indicators of this growing awareness emerged in the 1880s. The first was an article by E. L. Godkin, a journalist and newspaper editor who wrote in 1880 that a person's private life could be thought of as "that portion of the personality which is not physical or tangible, the tastes, habits, prejudices, sensitiveness, manners, relations with friends and family, and the like, about which the civilized man ordinarily dislikes to talk to strangers or have strangers talk." In Godkin's view, in modern life, a person's private life was always at risk:

> There never was a time when people did not enjoy hearing about their neighbor things which they knew he would not like to tell them. But ... we must admit that nothing is better worthy of legal protection than private life, or, in other words, the right of every man to keep his affairs to himself, and to decide for himself to what extent they shall be the subject of public observation ...

The community has a good deal to fear from what may be called excessive publicity, or rather from the loss by individuals of the right of privacy.[2]

The second was an 1881 Michigan court opinion addressing Alvira Roberts's complaint against a doctor who brought a stranger with him to assist with delivering her child. When Roberts went into labor and called Dr. De May, the physician brought a friend, Mr. Scattergood, with him. Roberts and her husband assumed that Scattergood, who remained inside their cramped, 14-by-16-foot home throughout her labor, was either a medical assistant or a doctor in training. According to court documents, when the Robertses later discovered that Mr. Scattergood was "a young unmarried man, a stranger to the plaintiff and utterly ignorant of the practice of medicine," they were mortified at the indignity caused by his presence, and they sued.

As it turned out, the Michigan court was equally appalled. The judges made a number of legal points in their opinion, but the most salient ones, from the standpoint of privacy, were these:

Dr. De May therefore took an unprofessional young unmarried man with him, introduced and permitted him to remain in the house of the plaintiff, when it was apparent that he could hear at least, if not see, all that was said and done . . . It would be shocking to our sense of right, justice and propriety to doubt [that] . . . the law would afford an ample remedy. To the plaintiff the occasion was a most sacred one and no one had a right to intrude.[3]

Although the court didn't use the word "privacy," its decision rested on what has become one of the thorniest principles of privacy: that a person can choose how widely to share information about themselves; that sharing it with one person doesn't create a license for it to be spread to all. (What makes this a thorny principle is the fact that it doesn't apply in all cases. Although it's an important cultural norm and a legal standard in some circumstances, we'll see in later chapters that the Fourth Amendment's "third party doctrine" takes the opposite approach.) Further, the Robertses could sue even though they had voluntarily allowed Scattergood into their home; De May hadn't fully informed them about who Scattergood was, and had De May informed them, they would not have given their consent.

The court acknowledged that this intrusion on privacy took place in the Roberts' home, a place that enjoyed long-standing, traditional protection. But the judges' language made clear that the Roberts' property interest in their home wasn't the only factor in their decision, nor even the most important one. Rather, the occasion was an intimate one. Scattergood could see and hear everything that happened during the childbirth, and De May and his friend hadn't fully informed the Robertses about the implication of their decision when they agreed to let Scattergood in. Although the court never used the phrase "reasonable expectation of privacy"—a concept that would later become important in privacy law—the judges strongly implied that *anyone* would find this set of facts to be "shocking to our sense of right, justice and propriety."

Although this case has long since been overtaken by other, more widely cited decisions, it is in some ways the birth—no pun intended—of the American justice system's recognition that when one person invades the privacy of another, the law should provide a remedy for that injury. This case laid the foundation for two key, enduring principles: that the law should recognize a right to privacy in certain situations, and that "consent" to an intrusion on that privacy is meaningless if the consent is given under false pretenses. As we'll see in later chapters, both of these concepts are, or should be, relevant to understanding data privacy today.

A few years after Godkin's article and *De May v. Roberts*, these nascent formulations took shape, and took hold, in an 1890 law review article, "The Right to Privacy,"[4] written by future US Supreme Court Justice Louis Brandeis along with his law partner, Samuel Warren. Warren hailed from a Boston society family and was incensed that Boston's paparazzi and local newspapers had intruded on his daughter's wedding. Irked, he and Brandeis wrote about how technology and the press were intruding on private life and heightening the need for individuals to enjoy a "right to be let alone." They wrote:

Instantaneous photographs and newspaper enterprise have invaded the sacred precincts of private and domestic life; and numerous mechanical devices threaten to make good the prediction that "what is whispered in the closet shall be proclaimed from the house-tops." For years there has been a feeling that the law must afford some remedy for the unauthorized circulation of portraits of private persons . . .

The press is overstepping in every direction the obvious bounds of propriety and of decency. Gossip is no longer the resource of the idle and of the vicious, but has become a trade, which is pursued with industry as well as effrontery. To satisfy a prurient taste the details of sexual relations are spread broadcast in the columns of the daily papers. To occupy the indolent, column upon column is filled with idle gossip, which can only be procured by intrusion upon the domestic circle.

Their description of the ways in which the technologies of the time made it possible for trolls to plant news and for gossip to go viral sounds familiar. With those words, Brandeis and Warren could as easily have been writing about the twenty-four-hour news cycle and the proliferation of social media in the twenty-first century. As an antidote to "the intensity and complexity of life," people needed "some retreat from the world." Yet, at a time when people needed solitude and privacy more than ever before, they lamented, cameras and mass-market journalism were subjecting modern man to "mental pain and distress, far greater than could be inflicted by mere bodily injury." As a result, they argued, the law must grant each individual "the right of determining, ordinarily, to what extent his thoughts, sentiments, and emotions shall be communicated to others."

In other words, Brandeis and Warren were advocating for privacy law that protected *people*, unrelated to their property.[5] The right to privacy, they wrote, must function

to protect those persons with whose affairs the community has no legitimate concern, from being dragged into an undesirable and undesired publicity and to protect all persons, whatsoever their position or station, from having matters which they may properly prefer to keep private, made public against their will.[6]

Building on the foundation laid by their article, US law came to recognize four distinct categories of invasion of privacy torts—that is, harms for which one person can sue another:[7]

1. Appropriation of name or likeness (using someone's name or image without their permission)[8]
2. False light (similar to defamation, saying or implying things about a person that aren't true and would be "highly offensive to a reasonable person")[9]
3. Public disclosure of private facts (spreading information—often embarrassing—that is factually accurate, but that the person involved did not want or intend to be widely known)[10]
4. Intrusion on seclusion (interfering with a person's solitude or "right to be left alone" in ways that would be "highly offensive to a reasonable person")[11]

It would take nearly seventy years for a similar theory of privacy to take hold in the law enforcement context. In a famous 1967 case called *Katz v. United States*, the court was asked to decide whether a bookie had a reasonable expectation of privacy in the illegal wagering calls that he was making from a public phone booth. In prior cases, the court had focused on property rights in determining whether police needed a warrant to carry out a particular search or seizure and had taken the view that, unlike opening a sealed letter, electronic surveillance wasn't a search or seizure at all.[12] In the *Katz* case, the court overruled its own precedents, holding that neither physical ownership of the premises nor physical trespass were the defining features of what the Constitution protected. When the government carries out electronic surveillance, like the wiretaps at issue in the *Katz* case, the court held that "the Fourth Amendment protects people, not property." (We'll see in later chapters that more recent Supreme Court cases have considered returning to a Fourth Amendment theory based on trespass to property rights; after several decades of relative clarity, the state of legal theory on the issue is once again in flux.)

Perhaps what's most noteworthy in this history is that there are striking similarities between the legal conundrums, social pressures, and driving factors that led privacy law to leap forward over a hundred years ago and the forces that are creating privacy risks and shaping privacy law today. A recognition that an individual should have the right to keep some things private; the impact of changing technology that makes it possible to record and disclose information about someone without their permission; even the nineteenth-century version of deep fakes—all of these issues were alive and well in the 1890s. In light of this, there is enduring wisdom in the conclusion Warren and Brandeis reached: "Political, social, and economic changes entail the recognition of new rights, and the common law, in its eternal youth, grows to meet the demands of society."[13] Put another way: technology changes; law and policy must evolve; and although it's intangible, privacy is a value that's essential to individual personhood and to a civilized society. Consequently, the law must protect privacy, whether it's being threatened by the government or private actors.

The lessons are clear. American legal theory advanced when courts and commentators were willing to take an innovative approach to address the challenges created by new technology. Alvira Roberts had a right to keep her childbirth private. Photographers were welcome to make use of their equipment to publicize people who wanted their image preserved, but companies were prohibited from using someone's likeness without their approval or consent. Above all, a modern and evolving society, a mature civilization, had the ability and the obligation to protect its citizens from the indignities of having their intimate details exposed. Whether it came in the form of anonymity in one's actions or the ability to keep one's thoughts, opinions, and preferences to oneself, the law was ready to step in and protect individuals from the intrusions made possible by new technology that exploited the worst human tendencies toward gossip, salaciousness, and reputational harm. As the *De May v. Roberts* court had held, "Where a wrong has been done another, the law gives a remedy."[14]

As smartphones, social media platforms, location tracking, deep fakes, and other forms of new technology have fundamentally reshaped what is knowable and known about us, the law has often struggled to keep up. But the privacy principles established in the nineteenth century were driven by social and technological change, and the principles were intended to endure. So while these principles have, to some degree, fallen out of currency, the heightened concerns about data privacy today make this an opportune time to revive privacy's history, to reflect on the utility of old ideas in a new age, and to adapt these principles to navigate the perils and promise of ever-advancing technology in modern life.

CHAPTER 4
WHAT'S IT TO YOU?

Understanding What Privacy Is Worth

S ocial scientists, behavioral economists, think tanks, and government researchers have undertaken studies attempting to measure what privacy is worth. The jury is still out, as it's proven to be a complex question, but the findings so far have been illuminating. First, how much people value privacy depends in part on their default position—that is, whether they have to pay to gain privacy or are being paid to give it up. Second, people sometimes value privacy differently in online transactions than they do when interacting face-to-face. And third, despite the fact that companies can make a handsome profit from our data, there is virtually no market existing today that allows individuals to directly profit from the information that might be considered "their own."

PAYING TO GET PRIVACY, AND BEING PAID TO GIVE IT UP

In an influential paper titled, "What Is Privacy Worth?" Carnegie Mellon University academic researcher Alessandro Acquisiti and his co-authors tried to measure the value of privacy. Their conclusion:

Individuals assign markedly different values to the privacy of their data depending on a) whether they consider the amount of money they would accept to disclose otherwise private information, or the amount of money they would pay to protect otherwise public information; and b) the order in which they consider different offers for that data.[1]

Acquisiti's work, in this and other papers, has laid the foundation for applying behavioral economics to questions about privacy. The research demonstrates some of the reasons why it's notoriously tricky to assign a monetary value to privacy.[2] First, data asymmetry is endemic: data subjects rarely know as much as data holders do about what's being collected and how it's being used. Second, data subjects seldom have complete visibility into, or a full appreciation of, the complex interactions among the many ways that data can be used. Third, even with that information and appreciation, consumers find their choices are limited. For all these reasons, people don't always make what appear to be fully rational choices when it comes to privacy.[3]

In addition to these challenges, society doesn't have agreed-upon yardsticks to measure the cost of privacy harms, and this lack of context leads individuals to make decisions on their own. The situation is different for many other kinds of harms, where the law and cultural norms give us well-established ways to calculate the injury, or "damages," that are involved. When someone is the victim of medical malpractice or injured in a car accident, when they're swindled in a business deal or wrongfully fired from their job, judges and juries can apply time-honored formulas for considering how much money those harms might be worth. When it comes to privacy, however, we don't have the same body of clear precedent or widely agreed-upon formulas. On the contrary, relatively few data privacy cases result in any financial compensation for the individuals whose privacy was intruded upon.

Recent years have brought a wave of data breaches involving credit card data held by department stores, Social Security numbers held by federal government agencies, and password information held by webmail providers, along with a range of other personal details. Yet many federal courts have held that plaintiffs in the resulting lawsuits don't have standing because even if millions of credit card numbers were stolen by cyber criminals, the breach of that data doesn't amount to enough of an immediate, tangible injury to the plaintiffs to support a claim for monetary damages in federal court. Given the reluctance of these judges to recognize standing for "traditional" data breaches involving credit cards and Social Security numbers, it's perhaps not surprising that the barrier to bringing a data privacy lawsuit is even higher when it comes to non-credit-related information like personal photos, biometric data, and password-protected online activity on social media sites, apps, and other web-based forums. Many people would consider those kinds of data to be more personal than a credit card number, and more like the "tastes, habits, prejudices, sensitiveness, manners, relations with friends and family, and the like," which Godkin argued it was so important to protect. Nonetheless, many courts haven't decided yet if they'll recognize an injury at all when that kind of information is misused, much less how much the resulting privacy harms might be worth.

Researchers like Acquisiti have discovered that one way to assess the value that people assign to privacy is to measure the gap between how much a person will accept in exchange for their personal data and how much they're willing to pay to keep it private. As the research shows, real-life privacy decisions can involve either kind of decision.[4] For example, we might demonstrate "willingness-to-pay" through accepting minor inconveniences like using search engines that don't track our history even if their search results aren't as complete as the market-dominating but privacy-intrusive results provided by Google search. Our willingness-to-pay could also take the form of increased out-of-pocket costs to enjoy more privacy, like subscribing to a paid online service that promises not to sell our information as opposed to a "free" version that isn't as privacy-protective. "Willingness-to-accept" decisions also abound in everyday life. In many instances, we make them by implication rather than deliberately, such as when we give up detailed personal information in exchange for free online services, from social media to using free tax preparation software or earning a pittance for completing online surveys.

To measure these differences, researchers offered mall shoppers two versions of a gift card: one worth $10 that didn't track the user's spending, and the other worth $12 that tracked how the card balance was spent. Shoppers with the $10 card were offered the chance to upgrade to a $12 card if they were willing to have their spending tracked. Most shoppers said no. When shoppers started with a default condition of privacy and asked what they would be willing to accept to give their privacy up, $2 wasn't enough.

For shoppers with the $12 card, the experiment's conditions were reversed. For these shoppers, the default condition was lack of privacy. When asked if they'd rather have private shopping on a $10 card instead, the majority of the shoppers declined. When it came to willingness-to-pay, it wasn't worth $2 to gain added privacy.[5] The shoppers reached opposite conclusions: for one group, privacy was more important than gaining an additional $2; for the other group, the $2 mattered more than their privacy. The shoppers had been selected randomly, and the findings underscored both the gap that can arise between willingness-to-accept and willingness-to-pay, and how much influence a person's default condition of privacy can have.

As Acquisiti and his colleagues point out, the decision in real life is rarely so stark as the ones presented in these experiments. As a result, it's very hard to gauge precisely what trade-offs between privacy, cost, convenience, and other factors people truly value the most.[6] When we're making real-life decisions about which smartphone to buy or which app to download, we likely consider a range of factors including cost, convenience, familiarity with the operating system and the device, availability of particular features, the opinion of a trusted friend, online reviews and ratings, privacy considerations—and more. How we weigh these factors is often complicated by having limited options. For search engines, social media, video streaming platforms, and other data-intensive services, there are often only one or two major providers in a particular niche, and a handful of minor competitors that might offer more privacy but whose service or functionality isn't as complete.

It begs the question: How meaningful are privacy choices, when the limited alternatives leave consumers to compare apples to oranges, weighing privacy features against other aspects of functionality, convenience, design, accessibility, and more? And how do we correct the asymmetry that leaves us all prone to undervaluing the potential privacy impact of the everyday decisions we make?

Just as most of us aren't single-issue voters when we step into the polling booth, most of us aren't single-issue consumers when it comes to making decisions about various products, services, and apps. Do we think about privacy? Generally, yes, to some degree, and some of us think about privacy more often or weigh its importance more heavily than others. But seldom do we make a purchase, download, or registration decision *solely* on the basis of privacy. The multifaceted nature of these decisions, coupled with the limited number of choices that we face in many instances—the available search engines, social media platforms, mobile phone operating systems, and device models—makes it difficult, no matter how well informed we are, to make judgments that assign a consistent value to privacy. We seldom face a straightforward calculation when deciding how much we're willing to accept to give up our privacy, or how much we're willing to pay to keep it.

The nation's leading economists also grapple on a macroeconomic level with measuring what privacy is worth. In an October 2019 speech, Federal Reserve Chairman Jerome Powell noted that traditional models of economic growth didn't seem to explain the level of gross domestic product in the United States when compared with other economic indicators.[7] Part of the explanation may lie in work being done by government and academic researchers trying to understand how free online services contribute to economic value and growth.[8] According to a team at MIT, when consumers were asked how much they would demand in exchange for giving up free apps and online services, the numbers were significant. The average American Facebook user, for example, spends twenty-two hours per week on the platform and would demand $48 to give it up for one month.[9] According to similar studies, mobile phone users in the European Union would demand €59 to give up navigation apps for a month, €97 to give up Facebook for a month, and €536 to give up WhatsApp for a month.[10] Giving up search engines was seen as the greatest sacrifice, with US users demanding $17,000 to forgo using them for one year.[11]

Although the Fed's research isn't aimed specifically at measuring privacy, it provides important insights into the value of free online services—services that are only "free" because the platforms can monetize their users' data. This exchange of personal data and the monetary value online services assign to it means that when measuring the value of data-intensive services, privacy is always part of the equation, even if it's unexpressed. Despite this inextricable link, there aren't yet many studies examining the intersection between no-fee services, the privacy cost (to the individual and society), and the economic value (to the individual, the company providing the service, and the economy and society as a whole). For example, in the MIT study, the researchers concluded that people who would demand large payments to give up free services must find value in those tools; therefore, free platforms and apps contribute to consumer well-being—yet the word *privacy* appears nowhere in their research.[12] It's clear that these research participants would want to be paid to give up data-intensive digital tools; but it isn't clear whether they know, or were asked if they care, about how much privacy they're currently giving up.

> *When measuring the value of data-intensive services, privacy is always part of the equation, even if it's unexpressed.*

This kind of research is expanding globally as countries around the world grapple with the privacy and other impacts of data-driven technologies. In 2019, the Australian Competition and Consumer Commission (ACCC) published its Digital Platforms Inquiry, one of the most comprehensive studies to date of the impact that large data platform providers have on individuals. The ACCC found that the business model of companies like Facebook and Google harms individuals in a number of important ways, allowing corporations to "exploit the information asymmetries and bargaining power imbalances between digital platforms and consumers."[13] That is to say, although each person may have different levels of risk tolerance when it comes to privacy, and different preferences when it comes to weighing privacy against cost or convenience, the digital platform model offers consumers very little opportunity to exercise choices related to those preferences.[14]

Although a great deal more research remains to be done, some striking patterns are clear. How much people value privacy depends in part on the amount of privacy they start with as their default. It also depends on how many options they have and how complex the trade-offs are that they have to weigh. People find value in data-driven online services, and it appears they are also swayed by how those services are marketed. That last factor is illustrated by the definition of "free" goods and services that's relied on by the nation's top consumer protection watchdog, the Federal Trade Commission. Under a 2012 FTC rule:

> The offer of "free" merchandise or service is a promotional device frequently used to attract customers. When making "free" or similar offers all of the terms and conditions upon which one can receive and retain the free item should be set forth clearly and conspicuously at the outset of the offer so as to leave no reasonable probability that the terms of the offer might be misunderstood.[15]

Currently, however, this rule hasn't yet resulted in a requirement that Google, YouTube, Facebook, Instagram, or other platforms explain to its users, in truly conspicuous terms, that "free" means "we're offering you the use of our service at no cost in exchange for mining—and perhaps selling to others—comprehensive data about what you have, what you do, who you are, and what you think, believe, and know." If the trade-off were presented that way, more people might pause before using some of these platforms and services and might create more market pressure for other options. Instead of demanding $17,000 to walk away from Google search, it might be that a little bit of inconvenience with less-tailored search results on a more-privacy-protective search engine is all that some people really want.

These experiments and research provide important insights into how our behavior reflects the true value we place on privacy. However, that isn't nearly the end of the analysis, especially as technology and data growth continue to accelerate. Much of American and European data protection law depends on the concept of informed consent. Consequently, it's important to ask what kinds of information an individual needs in order to be informed in today's data-rich environment—and whether the data ecosystem has become so complex, and the range of choices so limited, that it's no longer possible to give meaningful informed consent. A key challenge for the future will be how to clearly present people with useful information in an economic environment that offers meaningful alternatives in a manner that empowers them to consciously decide what privacy is worth to them, and then choose what trade-offs they are willing to make.

SECTION II

If You're Not Paying for the Product, You Are the Product

B y the summer of 2019, an estimated 2.4 billion people around the world had Facebook profiles, and nearly 1.6 billion people were logging into the social media platform every day. Years before, academic researchers and advertisers had discovered that analyzing a person's Facebook "likes" revealed a surprisingly detailed and accurate personal portrait.[1] In 2015, an academic journal published research showing that computer models equipped with a person's history of Facebook likes outperformed friends and family members in gauging the individual's personality traits and inclinations.[2] They studied 17,000 computer assessments and 14,000 human assessments, and concluded that with somewhere between 10 and 300 likes, the algorithms become more accurate than the humans in assessing someone else's personality. With 10 likes, Facebook knows you better than your coworkers do. With 70 likes, Facebook knows you better than your roommates and real-life friends do. With 150 likes, Facebook knows you better than your own family does. And with 300 likes, Facebook knows you better than your spouse.[3]

If companies can make individually tailored, and highly educated, guesses about us based on innocuous social media scrolling behavior, how should that shape our thinking on what limits should be in place for said companies to collect and use this information? Indeed, there is no clear, uniform, or clearly articulated legal framework for consumer data privacy in the United States. Companies that operate in the United States, from massive social media platforms to the tiniest home-basement-developed apps, have almost completely unfettered ability to collect, retain, compile, cross-reference, enhance, sell, share, buy, and use information about the consumer. In most cases, for most kinds of information, the only requirement is that they provide the user with a privacy notice explaining their data practices. The consumer may or may not have read or understood the notice, but they typically are deemed to have given consent if they keep using the service or if they've clicked on a box somewhere that says something like, "I accept." This notice-and-consent-based approach dates back nearly a half-century, and has taken root not only in the United States but in other countries around the world as well.

The 1970s and 1980s proved to be formative years in privacy-related technology, policy, and law. With the expansion of computerized record-keeping, policymakers in the United States were searching for the right balance of principles and practices that would facilitate expanding the potential societal benefit of online databases, while mitigating harm to individuals. Policymakers and thought leaders, such as the authors of the US government report that established the Fair Information Practices, believed that a framework of privacy notices coupled with consumer consent would allow the public as a whole to engage in reasoned review of government information practices.[4] National and international bodies looked to privacy notices and consumer consent as important tools to support individuals' ability to make decisions about the use of their data. By extension, this notice-and-consent model could provide some check on unsavory corporate behavior.[5] Now, decades on, it isn't hard to see that the effectiveness of the notice-and-consent model has been eroded by a combination of factors: the proliferation of wordy, unintelligible privacy policies; consumers' recognition that they're powerless to negotiate any better or different privacy terms; and the fact that new types of data tracking, collection, and analysis are being developed faster than consumers can become aware of them.[6]

In today's complex data privacy environment, it's more difficult for consumers to understand what they're consenting to: how their data might be collected and used by the owner of a free product or service they've signed up for, how it might be sold to others, what the impacts of cross-platform data aggregation are, and how artificial intelligence algorithms are creating behavioral prediction models about them. The reality is, those prediction models can be used for purposes as ordinary as direct marketing and targeted commercial advertising, as well as for purposes as consequential as political advertising or as sinister as political viewpoint manipulation by hostile foreign governments looking to sway public opinion in Western democracies.

When it comes to private-sector use of data, there is no equivalent to the Fourth Amendment protections that restrict government data collection. There is no federal data privacy law. And courts haven't decided how they feel about common law claims for invasion of privacy when a company takes, uses, shares, or loses an individual's data.

> *When it comes to private-sector use of data, there is no equivalent to the Fourth Amendment protections that restrict government data collection. There is no federal data privacy law. And courts haven't decided how they feel about common law claims for invasion of privacy when a company takes, uses, shares, or loses an individual's data.*

Faced with these challenges, state legislatures have been considering ways to fill the privacy protection gap. Illinois' Biometric Information Protection Act, passed more than a decade ago,[7] became a wellspring of litigation in 2019 when the Illinois Supreme Court held that individuals could file lawsuits against companies that collected handprints and other biometric information without their consent.[8] The lawsuits, ranging from the precedent-setting complaint against an amusement park's use of handprints for its ride lines to the now-routine lawsuits against companies who direct their employees to use fingerprints or handprints for logging in and out of biometrics-based company time clocks, have prompted other states to adopt similar laws restricting the collection and use of biometric data and have also prompted companies to lobby against these new proposals and urge changes in the Illinois law to take away the right of individuals to sue.[9]

In what has been the most significant change to US privacy law so far, California passed a sweeping new law, the California Consumer Privacy Act (CCPA), that took effect on January 1, 2020. The CCPA started as a grass-roots voter referendum intended to give individuals more insights into who is collecting "personal information" about them and why, and what they're doing with it. One of the hallmarks of this new law is that it vastly expands the kinds of data governed by the new privacy protections. Most state data privacy laws focus on breaches of narrow categories of information that typically center around Social Security numbers, payment card and financial account information, and sometimes medical records. Under CCPA, however, "personal information" is defined to include almost every imaginable fact or inference about California-based individuals or households: the definition includes traditional items like Social Security numbers and credit card and banking information, as well as IP addresses, online shopping and other internet activity, location data, biometrics, education and employment information, "audio, electronic, visual, thermal, olfactory, or similar information," and "inferences" that are drawn from personal data to create consumer profiles "reflecting the consumer's preferences, characteristics, psychological trends, preferences, predispositions, behavior, attitudes, intelligence, abilities, and aptitudes." Under CCPA, consumers can ask for access to, correction of, or deletion of their data, and can instruct companies that they don't want their data to be sold. Data collection about children under thirteen requires parent or guardian consent, and the law presumes that no one under the age of sixteen wants their data to be sold. The law only applies to companies doing business in California that meet certain revenue or data-handling thresholds, and the consumer protections only apply to residents of California.[10] By July 2020, enforcement had only just begun, and a long list of amendments were part of a ballot referendum in the November 2020 election. With so much uncertainty, it isn't clear yet what the law's long-term impact will be.[11] Nonetheless, given California's status as the world's fifth-largest economy, the law has a broad sweep, and privacy advocates in California are already working on changes to further strengthen the law. Meanwhile, legislators in other states as well as in Congress are watching CCPA developments to see what provisions might be adopted elsewhere and which ones are more trouble than they're worth.

Whether it's Facebook, Google, or the countless free apps and web-based services we use, the companies providing these services are getting something from us in return. Since we aren't paying them, they aren't getting money directly from us. Instead, they're monetizing our time on-screen. In some cases, they profit by offering us products from partners who pay them a commission when we make a purchase. In other cases, they profit by selling information about us to other companies who are interested in buying it. In still other cases, they create vast ecosystems of personal data about us and invite other companies to advertise to us in exchange for a fee. No matter what the specific business model is, the outcome is the same. When we're not paying for the product, we *are* the product. That doesn't mean that we shouldn't use these "free" services. But it helps, when we do, to know exactly what we're doing and to think about how much our privacy is worth.

CHAPTER 5
THE BIG 4

Apple, Google, Facebook, Amazon

O n April 21, 2016, a friend's post made it onto my Facebook feed:

> I may delete FB from my phone. Yesterday I used a gas station I'd never used before. An hour later an ad for that gas station appeared in my feed. My phone has become a totalitarian state and is spying on me.

Facebook—along with Apple, Google, and Amazon—knows you better than your own mother does. At least that's the conclusion of academic researchers and advertisers who pay to leverage the platforms' insights about you. It's even what these platforms' own marketing departments will say if you catch them in an unguarded moment. These four companies have unprecedented and unparalleled access to data about our preferences in news and entertainment, our online searches and purchases, our religious and political affiliations, our health, our education and employment, our hobbies and interests, our social connections, and our "psychosocial profiles." Their power and influence has grown so great that it's hard to recall their near-total dominance of entire data ecosystems is a recent phenomenon.

To understand the scope and reach of Apple, Google, Facebook, and Amazon, we should recall how much data has expanded in recent years and consider the ways that data matters to these companies. According to one report from December 2018:

> Ninety percent of the world's data was created in the last two years, and over 2.5 quintillion bytes of data are produced every day . . . And this data is then used to market products to us. In 2018, almost half of all advertising spend will be online, rising to over 50 percent by 2020. And two digital giants—Facebook and Google—now control 84 percent of the market. The companies are hugely reliant on ad revenue, with Facebook collecting 97 percent of their overall revenue from ad spending while at Google it accounts for 88 percent.[1]

For every minute in 2017, YouTube users watched over 4 million videos, Google responded to 3.6 million search requests, and Amazon earned over $250,000 in sales.[2] That's *every minute*—for an *entire year*. And every year, these numbers go up. Now think about the volume of personal information stored by just one of those companies, let alone all four. Every minute, Google collects and stores data about those millions of YouTube (they share a parent company, Alphabet) videos watched: which individuals are tuned in to which videos, who clicks on those videos served up in search results versus auto-load or recommender analytics pushing videos to the top of an individual's feed, and so on. Because of Google's market dominance in search and heavy market share in services like webmail and products like mobile phones, it can correlate the information about users' video-viewing habits with information about the identity of the contacts in their digital address book, what they're talking about in the emails they exchange, and where they're traveling using Waze or Google Maps—including the difference between their occasional trips and their daily routines. All of this data and more can be combined, correlated, and crunched with the data that Google receives and stores from the 3.6 million web searches it executes every minute. This is a staggering granularity of detail about individuals that has never been available to anyone—governments or corporations—in the past.

Perhaps it's because the rise of these companies is so recent that their privacy policies and practices have varied widely and changed frequently over the years. Government regulators in the United States and abroad have vacillated between wanting to encourage corporate growth and innovation and wondering when to rein them in. As governments around the world are beginning to grapple with the consequences of these vast data pools, most of the regulatory effort has been concentrated in regions like Europe and countries like Australia, Canada, and the United States. The greatest scrutiny has landed on the corporations—Facebook, Google, Amazon, and Apple—that have grown so large they function as extra-national fiefdoms, courting governments and challenging government mandates in every country where they operate. It isn't just privacy that's at stake; competition is suffering, too. Smaller competitors can't enter the market because of the dominant position of the digital behemoths. Startup companies wanting to introduce a new search engine point to Google's market dominance as a barrier to entry; new social media platforms point to Facebook and raise the same concerns. Within the United States, the most important curbs on corporate exercise of data-related power have come from the Federal Trade Commission and from private litigation, with a handful of states starting to step into the mix. As the nation's antitrust regulator and consumer protection watchdog, the FTC may be uniquely positioned to leverage existing legal tools to bear in considering whether these giants may have grown too powerful.

The first antitrust law in the United States, the Sherman Act, was passed in 1890 to break up the railroad, steel, oil, and sugar monopolies that were dominating the economy and politics of the late nineteenth century. By the time that Teddy Roosevelt took office, the federal government was aggressively pursuing antitrust litigation against a bevy of companies that were operating in "restraint of trade."[3] Early regulators focused primarily on the anti-competitive impact of vertically integrated companies, a trend that would continue for a century as regulators asked, "Are all the steps in a supply chain controlled by a single corporation? Are large companies preventing smaller ones from entering the market? Are consumers paying higher prices as a result?" Although FTC regulators investigating Standard Oil might not have anticipated a Google, Amazon, or Facebook, the principles captured in that last question will likely prove to be key to the ways that today's FTC thinks about the role that antitrust laws can or should play in data privacy: the idea that competition benefits consumers is what prompted the United States, and many other countries, to create a single regulatory agency charged with both anti-competition regulation and consumer protection.

Over a century of experience, however, indicates that merely measuring the price of products doesn't provide a complete picture of whether or how corporate mergers and growth are impacting individuals for the worse. As noted in the previous chapter, pricing alone doesn't capture the full spectrum of benefits that users enjoy in "free" apps and services. It also doesn't account for the nonmonetary costs, such as the ways in which technology intrudes on our sense of security, self, and autonomy; sets us up for viewpoint manipulation; and interferes with our right to be left alone. Perhaps it's no surprise, then, that progressive politicians are starting to join privacy advocates in arguing that big tech needs to be broken up.[4]

WHAT KINDS OF DATA ARE WE TALKING ABOUT?

Although many of us have a general sense that digital platforms collect a lot of data from us, it's often harder than one might expect to get a clear picture of precisely what data is being scooped up. It's even harder to understand how that information is used in drawing inferences about us and in attempts to influence us. The Australian Consumer and Competition Commission (ACCC) tried to tackle these questions in its 2019 report, explaining that platform providers go to great lengths to capture their users' attention: a longer attention span translates into more user data. To meet that goal, data is collected in three ways: actively (e.g., when a user enters their contact information in an online form, watches a video, clicks on a link, or navigates to a new page); passively (e.g., background collection of location data from Wi-Fi networks); and by inference (e.g., by analyzing active and passive user data to draw inferences about a user's age, gender, health, sexual orientation or identity, political affiliations, hobbies, interests, and so forth).[5]

The ACCC's research underscored the gap between the kinds of data that individuals consider to be "personal information" and the data that's covered by most privacy policies and data protection and data breach laws. According to the ACCC report, when Australian consumers were asked what kinds of data they viewed as "personal information," they included the following items in the list: date of birth (86 percent), a person's name (84 percent), photographs (79 percent), telephone and device information (79 percent), and location information (78 percent).[6] Under most US state data breach laws, name and date of birth are only considered personal information if they're combined with other, more sensitive information, like Social Security numbers or payment card information. Privacy laws generally don't protect location information, except with respect to certain kinds of government uses. Telephone and device information is largely unregulated by these laws. And under most state data breach laws, photos aren't protected at all.

We didn't need the ACCC report to see that gap; the platform providers' privacy policies are proof enough. By and large, privacy notices are written with an eye toward complying with whatever laws govern the data collection practices of the service, app, platform, or device. The risks and requirements under those laws are proliferating—back in 2012, researchers estimated that if a person were to read all of the privacy notices that accompanied every service they use, it would take seventy-six straight days to complete the reading.[7] That number is almost certainly higher now. Most of us largely accept the fiction that these privacy policies might impact the decisions we make. Even courts acknowledge that these privacy policies offer little more than a fig leaf of user notice and consent, since they are cumbersome to read, difficult to understand, and individuals have few alternatives when it comes to using the major digital platforms.

One stark example of this emerging view among courts came in a decision involving the class-action litigation filed against Facebook in 2018, alleging that the platform violated users' privacy rights when it shared personal information with the behavioral research and political marketing company Cambridge Analytica. The District Court for the Northern District of California had to assess, among other things, whether users consented to having their Facebook profile information, posts, photographs, and contacts' information shared with Cambridge Analytica so that it could target political messaging campaigns. Facebook's defense rested in part on the position that at least some of the provisions in the various versions of its privacy policy that were in effect at different times should have put users on notice that their profiles, and their friends' profiles, would be shared with third parties. As a result, Facebook argued, there was no harm and therefore no legally cognizable foul when Facebook allowed the consulting company to export information of some 87 million users. In addressing this issue, the court noted that, "The parties agree that California law requires the Court to pretend that users actually read Facebook's contractual language before clicking their acceptance, even though we all know virtually none of them did."[8] This creates a difficult conundrum for plaintiffs and judges. As Judge Vince Chhabria wrote:

> To be sure, for the rare person who actually read the contractual language, it would have been difficult to isolate and understand the pertinent language among all of Facebook's complicated disclosures. Thus, in reality, virtually no one "consented" in a layperson's sense to Facebook's dissemination of this information to app developers. But under California law, users must be deemed to have agreed to the language quoted [in the privacy policy].[9]

This particular opinion was written at an early stage in the litigation, and at the time this book was going to press, there had yet to be a final resolution. It also isn't clear yet whether other courts will give a sympathetic hearing to future claims from users who say, in effect, "Sure, there was a privacy policy, but I didn't read it because I knew I didn't really have a choice," or "I read it, but I didn't understand it," or "It changed so often that I couldn't keep up." On some of these points, Alvira Roberts, if she were alive today, might sympathize. Future courts may find a parallel among plaintiffs who were presented with a privacy policy whose language seemed to disclose what was being done with their data (the cyber privacy equivalent of, "My friend is here to assist you") while failing to draw attention to other crucial facts ("He's not a doctor or medical staff of any kind"). Until that happens, Judge Chhabria's opinion is noteworthy for its blunt assessment of the practical utility that privacy policies have for most users.

Platform privacy notices are frequently updated and changed, but a snapshot shows some typical definitions of personal information. Google's privacy policy, for example, defines personal information as "information that you provide to us which personally identifies you, such as your name, email address, or billing information, or other data that can reasonably be linked to such information by Google, such as information we associate with your Google Account." Facebook's policy doesn't include an explicit definition of personal information, but does refer to "information that personally identifies you" as "information such as your name or email address that by itself can be used to contact you or identifies who you are." Apple's privacy policy states that "personal information is data that can be used to identify or contact a single person."[10]

Many of the kinds of data that seem most "personal" to the average user—photos, videos, interests, and membership in closed groups—fall outside these definitions. Even worse, none of the definitions even hint at passive collection or inferences. These definitions don't tell their users that providers are tracking their location from their internet connection and picking up all manner of digital detritus that spills out of other, leaky apps on the user's device. Or that the provider's cookies are tracking the user's web browsing, online shopping, and more, even after the user logs out of the provider's platform or app. Or that the providers are running complex analytics across all of this actively and passively collected data in order to analyze the user's personality, derive inferences about their interests, or influence their future behavior. None of these definitions explain that many platforms share personal data with corporate partners like data brokers who may sell the information to still other corporate partners (see sidebar on page 76), or that the platforms

USER ENGAGEMENT TACTICS: COMPANIES KEEPING YOUR EYES ON THE SCREEN

This is why YouTube and Netflix automatically cue up the next video as soon as the one you're watching is almost done, or why they make you wait until the next video has fully commenced before you can pause or stop the automated queue. It's because studies have shown that when a video simply ends, people are less likely to start a new one. Auto-queuing the next episode in a series lures people into binge-watching.

Just like willingness-to-accept and willingness-to-pay are often driven by the default privacy setting, how long people stay online—and provide more data for apps and platforms—is often driven by whether the default is an infinite scroll or an infinite video queue. Proactive steps always take more effort. So video games require half a dozen separate clicks to exit the game, and platforms make the default setting one that allows passive users to keep watching or scrolling into eternity.[11] According to a growing body of insider accounts, research, and reports, "Features such as app notifications, autoplay—even 'likes' and messages that self-destruct—are scientifically proved to compel us to watch/check in/respond *right now* or feel that we're missing something new or important."[12]

invite corporate partners into their ecosystems to support paid, microtargeted advertising. And none of the blandly worded policy notices advise users that, when platforms say they may share user data with their "partners," the scope of that sharing could include hundreds or thousands of entities and individuals, including app developers who pull in detailed personal information through application programming interfaces (APIs).

When Facebook, Google, and other companies allow third-party developers to provide new products and services on their platforms, those new apps encourage users to spend more time on the platform. This increases the advertising value of the platform and gives both the platform and those app developers more data about the user, which can be used to further target those users and shape their behavior. This vicious cycle is so opaque that many users don't know it exists. But the cycle is that way by design—it begins with teams of psychologists and social scientists who educate Silicon Valley on how to change user interfaces and services in order to entice people to stay online longer.

Lest individuals should think they're immune to this kind of data gathering so long as they don't use the major services that carry out these practices, the reality of the situation is heavily lopsided, and not in individuals' favor. Google, for example, collects information not only from individuals' own use of Google services, such as Gmail, Google's Chrome browser, and Google search, but also via the visiting of websites that use Google's analytics or advertising services. Similarly, Facebook tracks data of users who are logged into their Facebook accounts, users who are logged *out* of their Facebook accounts, and individuals who don't even have a Facebook account but who visit pages that have Facebook "like" buttons or other Facebook plug-ins that let the website owner boost traffic to their page or take advantage in other ways of the analytics and advertising opportunities that are made possible by Facebook's global reach.[13]

Although consumers and regulators alike are becoming more aware of the kinds of information being collected through digital means, rapid changes mean there are always new surprises. In February 2019, reports surfaced that Facebook was harvesting sensitive health information from people who were using completely independent apps.[14] The mechanism was a simple one: Facebook created a tool called App Events. When non-Facebook apps built App Events into their design, App Events allowed Facebook to gather data from those apps. Some data was relatively benign, like how many times a day the app was opened, or how long it was used during a particular session. But App Events could also gather data that the user manually input. In the case of health and fitness apps, this included things like blood pressure, weight, medications, exercise, heart rate, blood sugar, and the like. What grabbed the headlines, however, was the fact that, based on this feature, apps like Flo Health were sending Facebook data about the timing and frequency of women's menstrual cycles and sexual activity, and the apps' users had no idea—and no intention—of giving that very personal information to Facebook.[15]

WHAT KINDS OF CONSENT ARE WE GIVING?

By and large, the consents that we're giving are blanket, wide-reaching, and uninformed. That's a stark, bleak description of the situation, but it also provides a fair thumbnail sketch of the ways in which privacy notices and consent operate for most of the major digital platform providers. The advantage lies entirely with the platform. As one pair of researchers noted, "When a company can design an environment from scratch, track consumer behavior in that environment, and change the conditions throughout that environment based on what the firm observes, the possibilities to manipulate are legion." With that kind of reach and influence, the platform is able to "reach consumers at their most vulnerable, nudge them into overconsumption, and charge each consumer the maximum amount that he or she may be willing to pay."[16]

WHAT'S BEING DONE WITH OUR INFORMATION?

Once it's been harvested, our data takes on a life of its own.

Sometimes, our data sits right where it started. In 2019, Apple famously rolled out an advertising campaign with billboards proclaiming that "What happens on your iPhone, stays on your iPhone." It was a catchy slogan, and well designed to capitalize on Apple's carefully cultivated reputation for privacy. The campaign was timed perfectly to coincide with a major electronics industry conference in Las Vegas, where the ad's theme echoed the old and slightly scandal-suggesting trope that "What happens in Vegas, stays in Vegas."

Within the community of information security and privacy researchers, the billboard campaign was met almost immediately with a sentiment of, "Challenge accepted." Sure enough, it didn't take long for researchers to discover that the marketing hype was, to a large degree, merely hype.

In May 2019, the *Washington Post* reported that it had carried out tests to see just how much data really did stay on the iPhone. *Post* technology columnist Geoffrey A. Fowler teamed up with a security research lab to assess his iPhone's activity and found that, within seven days, the phone had exported data via 5,400 hidden app trackers.[17] Fowler's location information, IP address, phone number, and other device-identifying information were being exported off the device to thirsty apps that slurped the data in. The security research lab estimated that the trackers would have exported 1.5 GB of data over the span of a month—an amount that could easily chew up half of the monthly allotment for someone subscribed to a basic-level phone plan.[18]

Although Apple might be the emperor of privacy-based marketing campaigns, it wouldn't be fair to say the emperor has *no* clothes. After all, Apple devices leak less data than equivalent Android-based phones that leverage Google's vast data empire. But Apple might only be wearing a tank top and shorts, and not the full three-piece suit they've led us to believe.

Facebook and Google, by contrast, don't make sweeping promises about protecting user data. But they also don't generally sell it outright, and they don't sell it to third parties as often as people might assume. Instead, Facebook and Google operate "walled gardens of data."[19] That is, they create an ecosystem within which personal data is generated by users who intentionally volunteer information such as name, email address, birthday, searches, interests, photos, videos, contacts, and the like. And users add to that growing garden of data with information they provide unintentionally. Indeed, every time a user logs into Facebook or carries out a search on Chrome or sends or receives mail via their Gmail accounts or uses their Android phone, their information is collected by the gardener, or ecosystem host, and added to the already-rich profile of information on hand. Data about activity inside the garden is supplemented with data about activity outside the garden—activity that the gardener can track by means of the footprints that we leave as we traipse around the internet and stumble across the piles of mud that the gardeners have left to trip us up outside of their walled ecosystems.

DATA BROKERS: THE INVISIBLE BUSINESSES THAT SEE US ALL

Sometimes, our data takes flight, traveling in nanoseconds around the globe and through the servers and algorithms of companies whose names we've never heard. Our data gets shared with data brokers and aggregators who further compile, collate, collect, massage, integrate, interpret, analyze, sell, and re-sell our information to a range of bidders in a marketplace that we have little insight into and virtually no control over.

Broadly speaking, "data brokers" are companies, or business units within companies, that earn their primary revenue by supplying data or inferences about people that are gathered mainly from sources other than the data subjects themselves.[20] Because brokers generally get their data from third parties, consumers are largely unaware of their activities or of the profiles that data brokers are maintaining on them. In order to monetize these profiles, data brokers frequently create lists of people with shared attributes. In some cases, the lists seem tied to relatively benign, nonsensitive information: dog owners, winter activity enthusiasts, or "mail order responders."[21] In other cases, the lists relate to medical conditions, like wheelchair users, people with cancer, insulin-dependent diabetics, or people with depression. They may be sorted to identify people with breast cancer, impotence, vaginal infections, or HIV. Other lists are tied to religion, ethnicity, immigration status, or national origin, while others still are linked to economic circumstances, such as "Pay Day Loan Central—Hispanic," "One Hour Cash," or "Help Needed—I Am 90 Days Behind with Bills." Lists may also reflect family status, such as "expectant parent"; a combination of socioeconomic and family status factors, such as "upper-middle class with no children"; or shorthand categories created by the data brokers themselves, such as "rural everlasting" to refer to single men and women over the age of sixty-six with "low educational attainment and low net worth." In some cases, data characteristics are aggregated by neighborhoods, buildings, or households; in other instances, the data is specifically tied to identifiable individuals, with each person in a household uniquely identified for purposes of the data profiles and their membership on various lists.[22]

How do data brokers make money off of all this information? Generally speaking, they sell (or rent) personal data to other companies who want to use it for three main purposes: marketing, risk mitigation, and people search.[23] In the marketing context, the lists and individual profiles are used to analyze, segment, and sort prospective customers for targeted advertising campaigns based on particular characteristics, behavior, profitability, and projected lifetime value as a consumer for the company.[24] Companies interested in risk mitigation products are usually looking to confirm individuals' identities or detect fraud. "People search" services are often available online to any user, with limited information returned at no direct cost to the searcher, and more comprehensive profiles available for a fee.[25]

In 2014, the US Federal Trade Commission renewed its calls for Congress to pass legislation governing the activities of data brokers. The FTC acknowledged that the data broker business can provide benefits to consumers by increasing the likelihood that consumers will be presented with ads for products and services they're interested in. However, the FTC also pointed out the risks to privacy and data security of having all of this consumer information held in just a few hands, and recommended that new laws be passed that would, among other things, give consumers the right to find out what information data brokers have about them and to opt out of having their data sold for marketing purposes. Thus far, Congress hasn't made any progress in this area. But states are beginning to take action, with Vermont passing a data broker registration law in 2018 that requires companies whose primary business is the aggregation and sale of personal data to register with the Vermont attorney general.[26] And, as noted, the California Consumer Privacy Act of 2018 restricts the sale of data about children, gives consumers the right to prevent companies from selling their data, and requires that data brokers release detailed statistics about the types and volume of the data they collect.[27]

For example, Facebook tracks users outside their platform by embedding Facebook tracking pixels on participating websites. These pixels are invisible to users, and allow Facebook to track their activity on those sites *regardless of whether the site visitor has a Facebook account.* Facebook employs other external tracking tools as well, such as the Facebook "like" button that appears on many websites, the "login with Facebook" function available for many other platforms, and Facebook analytics, which many websites use for measuring traffic to and through their sites. As of April 2018, the Facebook "like" button appeared on 8.4 million websites, the Facebook "Share" button appeared on 93,000 websites covering 275 million web pages, and there were 2.2 million Facebook pixels installed on websites around the world.[28]

Although major platforms such as Facebook and Google don't generally sell user data to advertisers, they do make that information directly available to third-party app developers. Between February and April 2018, there were approximately 1.8 million apps on Facebook and 1.5 million app developers active on Facebook. Although the apps and developers are supposed to be operating within the confines of Facebook's privacy policy, that policy allows apps to create their own privacy notices that users seldom read and never consider objecting to. Facebook insists that it has remedied the practices that allowed third-party app developer Cambridge Analytica to siphon off the detailed personal information of some 87 million users who had never given consent to sharing their information. In July 2019, the US Federal Trade Commission levied a fine of $5 billion on Facebook because of its data handling practices, included the broken promises that were demonstrated by the Cambridge Analytica scandal.[29] However, many commentators pointed out that $5 billion, although precedent-setting in its size as it dwarfed the previous-largest privacy-related fine levied by a US regulator, was still only a fraction of Facebook's quarterly profits and cash reserves, amounting to little more than a slap on the wrist.[30]

In addition to that wide world of data sharing, the fact that companies of all sizes and across all market sectors and industries have access to so much information about us makes it possible for corporate and government entities to make a whole range of data-driven assumptions, conclusions, and decisions about individuals, including decisions that are inaccurate, arbitrary, or biased.[31]

DO PRIVACY AND COMPETITION INTERESTS ALIGN?

By 2017, antitrust regulators in the European Union were considering action against tech giants like The Big 4. The EU Competition Commissioner expressed concern about the ways in which the platform providers' data advantage served as a barrier to entry for other businesses. Since then, some countries in Europe have stepped out in front. For example, in 2019, Germany's Federal Cartel Office ruled that when Facebook harvests and processes data from third-party sites, it violates European data protection law, because users had no meaningful opportunity to object to the third-party data collection; their only choice was between widespread data collection or not using Facebook at all.[32]

Meanwhile, in the United States, Senator and 2020 presidential candidate Elizabeth Warren (D-MA) proposed a plan to "break up big tech," emphasizing that undoing some of the massive mergers of recent years—Amazon's purchase of Whole Foods and Zappos, Facebook's purchase of WhatsApp and Instagram, Google's purchase of Waze, Nest, and DoubleClick—could have the benefit of making tech companies more responsive to users' concerns, including those about data privacy.

The US Federal Trade Commission has been soliciting input on whether the biggest tech platforms are engaging in unlawful anti-competitive practices, and whether antitrust enforcement action could have ancillary benefits for privacy. In addition to the potential for federal-level action, investigation, regulation, or enforcement from the FTC, all fifty US states have consumer protection laws, many of which are closely modeled on the national FTC Act.[33] As the FTC has been exploring antitrust implications of big tech, it has invited input from states, where new litigation and legislation sometimes move more quickly than at the federal level.

As part of that federal-state interaction, in October 2018 a dozen attorneys general, representing eleven states plus the District of Columbia, wrote to the FTC about their concerns over data privacy and competition.[34] In their letter, the state AGs point to the central trade-off that these companies rely on: users' willingness to "make certain of their personal data available for monetization in return for the often 'free' services they receive."[35] The AGs were concerned that so much data is concentrated in the hands of just a few companies: nearly all searches use just one search engine, "over 90 percent of young people have a profile on one social media platform," and 99 percent of smartphones use either Apple's iOS or Google's Android operating system.[36]

According to the AGs, large-scale data aggregation by a small number of platforms can lead to a number of anti-competitive harms.[37] First, consumers suffer: the "immense" power imbalance between market-dominating platforms and consumers results in lengthy and opaque user agreements and few realistic alternatives for consumers, setting up a cycle in which consumers believe they have no choice but to agree to the platforms' collection of their data. Second, the big platforms' data advantage chokes out competition: without deep and detailed data about individual consumers, rivals are unable to serve up equally targeted advertisements, and therefore unable to attract the advertising dollars they need to stay afloat. To illustrate the problem, the letter noted that having access to historical search data improves the quality of new search results by up to 31 percent. "In effect," the AGs wrote, "today's search engines cannot reach high-quality results without this historical user behavior."[38] (Ironically, one counterbalance to this problem might be to allow more companies to have access to consumers' historical information—which could result in better search results but could also undermine privacy by making the data available to a wider base of companies.)

The AGs also urged caution with respect to big data algorithms, noting that in some cases algorithms could lead to price-discrimination or price-targeting that disadvantaged certain groups—a risk made more acute by the fact that there's so little transparency around how algorithms reach their conclusions.[39] (Issues surrounding algorithmic analysis and decision-making are addressed in greater depth in chapter 7.)

Perhaps the most important point was the AGs' position that "focusing on price to consumers is too narrow an interpretation of the principles of antitrust law."[40] With this statement, the AGs opened the door to considering privacy, discrimination, and other impacts from data-driven technologies—effectively undermining the ability of platforms, apps, and services to defend themselves solely on the basis that their products are "free."

The scope of the 2019 FTC fine against Facebook illustrates the ways in which a handful of major tech platforms have become so large that they're virtually ungovernable. When privacy advocates protested that $5 billion was far too little for Facebook to pay for privacy violations spanning nearly a decade, part of their rationale was that, in the same quarter that the fine was announced, Facebook earned $15 billion in revenue, and the company was sitting on $40 billion in cash reserves. Facebook's global userbase and deep pockets reinforce the power of its walled garden of data, further cementing its monopoly position and continuing to create incentives for millions of unaffiliated websites to embed Facebook tracking mechanisms.

Where the tech sector had enjoyed decades as the darlings of American economic innovation, federal and state regulators are now starting to question whether it's wise to allow so much of that growth to take place without some degree of oversight or regulation. In October 2019, a bipartisan group of forty-seven attorneys general launched an antitrust investigation into Facebook. Their individual press releases offered statements as varied as their constituencies: couched in different terms, with different areas of emphasis. Nonetheless, they agreed on a joint statement indicating that they "all are concerned that Facebook may have put consumer data at risk, reduced the quality of consumers' choices, and increased the price of advertising." All committed to "use every investigative tool at our disposal" to investigate the social media behemoth.[41] A bipartisan, multi-state investigation of this scope and scale would have been virtually unimaginable five years ago. That it's gained such widespread national traction is evidence of the growing realization that consumers may suffer as much or more from the ways that anti-competitive practices undermine their privacy as they do from practices that increase their out-of-pocket costs for participating in modern society and in the digital economy.

CHAPTER 6

WHEN YOUR DATA GOES TO SOMEONE YOU DIDN'T EXPECT

I n January 2017, two announcements came within days of each other: the US Federal Trade Commission issued its staff report on cross-device tracking, and Google announced that it would make all of its user search history data available to YouTube, so that YouTube could customize its identity-based advertising.

It's worth noting here that identity-based advertising isn't at all the same as identity politics—although it is in some respects a cousin. Identity-based politics, loosely described, is the approach through which politicians and their parties attempt to drum up support by appealing to people's sense of *who* they are: by pointing to demographics and trying to predict the likely actions of voting blocks—the elderly vote, the soccer mom vote, the African American vote, the NASCAR vote, and so on. Demographic groups are sliced and diced in countless ways, all with the intention of figuring out what people want, how to influence their views, and how to gain their support.

This identity-based approach is much the same as what marketing and advertising companies did in previous eras on Madison Avenue. They identified groups of consumers with similar characteristics, tried to intuit their fears, desires, and needs, and then looked for creative and aggressive opportunities to serve up messages that would appeal to those tastes. Identity advertising at the demographic level has been on display for decades: even children, if they're watching tv, are aware of the differences between the advertisements that air during *SpongeBob SquarePants* and the ones that come on when their older siblings watch *The Simpsons*.

Identity advertising in the modern era is far more granular than that, delivering ads to particular devices and accounts based on their pattern of use. Parents who share devices with their children are all too aware of the problem. Parents might use those devices at night, after the kids have gone to bed, to search for items that, in an adult context, may be private but aren't inappropriate: lingerie or sex toys, the best price on Viagra, articles on partner intimacy. They wouldn't want ads for those getting displayed to their kids, but that's exactly what happens when the kids pick up the same device, logged into the parents' account, the next day.

In an interesting twist on privacy, research has shown that many adults would rather use online services for "intimate" searches and purchases, because they find it less embarrassing to order condoms online than make a face-to-face transaction from the human cashier at the corner store. The online browsing experience might also be more pleasant and offer a wider selection; depending on local zoning laws, shoppers might find themselves resorting to seedy neighborhoods in order to find adult items in person. But the main reason so many adults gravitate toward digital options for consuming pornography, reading romance novels with steamy sex scenes, and responding to sex-related polls and quizzes is that they feel anonymous when they do these things online. On the pornographic video-sharing website Pornhub, for example, users answer questionnaires about their tastes and interests so that a "PornIQ algorithm" can recommend clips. Sometimes the online data reveals topics that people would find uncomfortable, or socially taboo, offline. Internet users in southern Bible Belt states, for example, ask more online questions about homosexuality and search for gay porn more often than in other parts of the country, even though the percentage of openly gay people is lower in the Bible Belt than the national average. Conversely, states that have higher percentages of openly gay residents make fewer online searches along those lines, even though it would likely be more socially acceptable in those more diverse states. When the Fifty Shades of Grey series of novels hit the market, six times more copies were sold on e-readers than in paper form, raising the distinct possibility that readers preferred the e-book format because it spared them the potential embarrassment of having family, coworkers, or neighbors see what content they were reading.[1]

Research shows that people value privacy in different ways.

Research shows that people value privacy in different ways. Customers shopping online have to surrender a lot of details about themselves: name, address, payment card or other financial account information, email address—perhaps an affirmation that they're older than a certain age if they're buying sex toys, for example. But what they don't have to do is look anyone in the eye, or face someone who might scowl at them or make awkward jokes, or start wondering what exactly they plan to do with those toys, and who they plan to do it with. On a rational level, the in-person purchase would seem to offer greater privacy protection: the purchase could be made with cash in a store that the customer seldom visits—they could even make the purchase in another town or another state. Depending on the decisions they make about where to shop, the chances of running into someone they know personally are small. There would be no permanent record of this purchase tied to them; it would give their transaction long-term anonymity. Yet many people would rather avoid the direct human interaction—even with a stranger behind a cash register—when buying things that feel intimately personal.

It would be condescending to those consumers to say that it's "ironic" or counterintuitive that they prefer online shopping over in-person interaction. After all, these shoppers are well aware that they're providing name, mailing address, email, and so on to the online retailer. They know their credit card company tracks the purchase, and they know the mail carrier might see a return address that isn't entirely discreet. If they're paying any attention at all, they know that their online searches are shaping the kinds of ads that are blotting out the screen space on their laptops or tablets or smartphones. So it isn't that consumers don't know that their intimate purchase information is being harvested and monetized and associated with their online user profiles; it's that, when it comes to sex and intimacy, they care about privacy in a different way. Where privacy advocates often focus on the data that's collected and retained, human behavior shows that, in some instances, people place more value on avoiding awkward face-to-face interactions than on limiting what goes into their digital profiles.

Which brings us back to identity-based advertising. This advertising is informed by our "online personas," the profiles associated with individual user accounts and combined with externally available information about our lives to extrapolate conclusions about us and our households. For example, a family of four might have an Amazon Prime account and may use Amazon for a range of purchases from books to groceries to clothing to electronics. It's a safe bet that each of the four people in that family will have different personal interests: different hobbies, different shopping habits, perhaps even different political or religious views. The online persona of that household, based solely on its Amazon profile, could look somewhat schizophrenic, with death metal music being streamed alongside gospel or country and western or Broadway show tunes. Not surprisingly, the same thing is true when it comes to online search parameters in Google or YouTube or any of the other major search engines or content delivery platforms. Although many devices support multiple user profiles, human nature often tends toward the path of least resistance, particularly in matters that don't strike us as significant. So it isn't uncommon for users sharing a device to simply search in whatever window is open, without first checking to see if another user's Facebook or Gmail account is running in the background (which would link the searches to that Facebook user or Gmail username).

This cross-platform tracking, coupled with the ad auction ecosystem to be discussed later in this chapter, has made it possible for advertisers to find prospective customers for their products and services with almost surgical precision. Where traditional advertising relied on mailing catalogues and flyers to homes based on zip code, or placing billboards in certain parts of town, or running television spots during certain time slots or on certain shows, those older models were based on broad generalizations about the demographics of a zip code, a neighborhood, or a viewing audience. Advertisers used to rely on industry surveys to tell them roughly who they might reach: consumers within a certain age range, or of a preferred gender, or in a particular earning bracket. But the information often didn't get a great deal more granular than that, until marketers had the ability to start combining wide-ranging demographics with information from activities like store loyalty programs. Even then, it was often far too difficult to identify prospective purchasers at an individual level.

The advent of the internet and cross-platform tracking, along with constantly growing data in the hands of data brokers, proved to be game changers for the advertising business. Instead of searching out places and hoping, based on the odds, that they might find prospective customers there, now advertisers have the ability to identify specific, individual customers, pinpoint their "location" on the internet, and go to where those customers are, even if the customer's online activity is showing up in a highly unusual place.

THE EVOLUTION OF ADTECH, OR HOW EVERYONE SEEMS TO KNOW WHAT YOU WANT

By the mid-1990s, the New York advertising industry realized that the internet could reshape the advertising business and, in 1996, formed a nonprofit trade organization: the Interactive Advertising Bureau. Now operating in forty countries, the IAB is the chief lobbyist and standard-setting body for the digital advertising industry. Its members include "media companies, brands, and the technology firms responsible for selling, delivering, and optimizing digital ad marketing campaigns." The IAB describes its mission as a proactive, pro-industry one: it "empowers the media and marketing industries to thrive in the digital economy."[2]

In its standard-setting role, the IAB created, and maintains, the Open Real-Time Bidding (OpenRTB) framework for online ad auctions. This is one of the main channels through which individually targeted advertisements are served up to internet users.

In general terms, the online advertising auction process works like this:[3]

1. You navigate to a website. As you click on the link, the website creates a "bid request" to send to an adtech channel, such as OpenRTB.
2. In this bid request, the website provides information about you: your IP address, the kind of device you're using, any personal details from previously collected data (like browser cookies showing what other sites you've visited), and behavioral profile information they may have bought from a data broker.
3. All of that information is bundled into the bid request and sent to an online auction where advertisers bid for the chance to show you their content. The winning bidder places their ad on the page you requested—all before the page is loaded into your browser window.
4. The entire process takes place in about 100 milliseconds—less than the time it takes to blink.

The IAB lobbies governments around the world to allow the digital advertising economy to continue to thrive. It keeps an eye out for proposed privacy laws that could inhibit digital advertising. It makes recommendations, issues policy papers, and carries out all of the other activities typical of trade associations that are interested in maintaining or growing their sector of the economy.

Meanwhile, the data brokers and aggregators whose activities had previously gone largely unregulated began drawing attention as the European Union passed the 1995 Data Protection Directive and the 2016 General Data Protection Regulation. Companies like Acxiom, whose entire business model depends on collecting bits of information and assembling them into detailed profiles of individuals around the globe, had in many respects flown under the radar when it came to data privacy regulation and enforcement. Laws in the United States and around the world allowed all manner of activity to be carried out with information so long as the individual who the data was from, or about, had consented to its collection, analysis, and sharing. While consent would remain an important and valid legal basis for data collection under the new European law, the GDPR required companies collecting information to provide more detailed, clear, and comprehensible explanations to data subjects, and to make sure that they provided proactive, unambiguous informed consent prior to the data collection. It also required data brokers and aggregators to respond to requests from individuals who wanted to know what information about them had been collected, and to correct or delete it at the individual's request.

In late 2018, a UK-based advocacy group, Privacy International, filed a complaint with the data protection authorities in England, Ireland, and France against seven companies, including data brokers Oracle and Acxiom, credit reference companies Experian and Equifax, and adtech companies Quantcast, Tapad, and Criteo.[4] The scope of data gathering and profile creation by these companies is hard to fathom: Acxiom, for example, has detailed data profiles on more than 700 million people, sorting them into neat buckets of potential interests that can be used by advertisers for marketing and sales campaigns.

On its website, Privacy International explains that among its many concerns about these companies is the fact that because, by and large, they're not consumer-facing, people don't know what the companies do or how to challenge them.[5]

DATA AGGREGATION AND THE SHARING ECONOMY

When it comes to the data sharing that fuels the gig and sharing economy, there's an important difference between local produce co-ops and ride-sharing bulletin boards, and multibillion-dollar companies like Uber and Airbnb. Many of those differences extend beyond data privacy. They include broader economic issues like discrimination in providing transportation and lodging; competition between licensed taxi and hotel providers and freelance drivers and room-renters; and the impact on road traffic and increases in housing prices in areas where ride-hailing apps and overnight-home-rental services like Airbnb are widely available. Those other issues fall outside the scope of this book. But the sharing economy, particularly in its big corporate incarnation, has had a significant impact on personal data privacy as well. As one set of researchers put it:

> Platforms like Airbnb, Lyft, and Uber possess deeply asymmetric information about and power over consumers and other participants in the sharing economy. And they are beginning to leverage that power in problematic ways. The sharing economy seems poised to do a great deal of taking—extracting more and more value from participants while continuing to enjoy the veneer of a disruptive, socially minded enterprise.[6]

The upshot? The researchers say, "Today's companies relentlessly study consumer behavior and use what they discover to maximize their bottom line." According to many observers, digital platforms are even better at doing that—and more insidiously poised to—than traditional marketing and advertising endeavors. The very fact that consumers are providing information to a digital platform creates an unfair advantage:

> When a company can design an environment from scratch, track consumer behavior in that environment, and change the conditions throughout that environment based on what the firm observes, the possibilities to manipulate are legion. Companies can reach consumers at their most vulnerable, nudge them into overconsumption, and charge each consumer the most she may be willing to pay.[7]

In the sharing economy, platform companies like Uber, Lyft, Airbnb, and others do more than simply facilitate connections between individuals; they also collect data about those individuals as the price for using the platform.

Uber, for example, reportedly uses a tool called Greyball to identify the phones of city officials, code inspectors, and law enforcement officers and purposely makes it difficult for those officers to find Uber drivers and issue them citations, effectively creating a "fake version of the app, populated with ghost cars."[8] According to other accounts, Uber regularly displays misleading information, showing a fleet of nearby "phantom cars" to make it look as though plenty of Uber rides are available when in fact the customer might be better off taking a taxi. As soon as the customer clicks to request a ride, the "phantom cars" disappear and the consumer faces a wait time longer than they anticipated, perhaps longer than a taxi might have taken. Uber reportedly manipulates its drivers as well.[9] Because Uber tracks the location, speed, and activities of its drivers on a near-continuous basis, it may be using the drivers' behavior to create and gather the data that will help it put those drivers out of work—either by contracting with other drivers who will do the same work for lower rates or through the deployment of a fleet of automated, driverless cars.[10]

The sharing economy platforms often lull consumers into a false sense of security in which they often aren't aware, or don't actively consider, the fact that they are paying in more than just money.

The sharing economy platforms often lull consumers into a false sense of security in which they often aren't aware, or don't actively consider, the fact that they are paying in more than just money. Because consumers are paying for the service, they often don't consciously consider the reality that they are also providing detailed personal information that these sharing platforms may use to exploit them.[11] Three use cases help illustrate this problem. First, Uber's data studies show that riders are willing to pay more for an Uber when their mobile phone battery is running low. Although Uber says it doesn't artificially inflate prices based on a user's phone battery charge, one has to wonder why else Uber has designed an app that measures a phone's battery status *and* done research about riders' tolerance for higher prices based on the level of their battery charge.[12]

The second example involves Uber's formula for calculating surge pricing—a strategy in which businesses can set flexible and dynamic prices based on supply and demand, competitor pricing, and other external factors. Surge pricing explains the higher costs for rides during peak times and locations, like near major concerts or sporting events. Uber's data scientists have found that consumers are wary of flat multiples (e.g., two times the usual rate for a particular route) but less likely to view fractional multiples as price gouging. With this information, Uber can set its surge pricing in ways that appeal to rider psychology, dissuading riders from turning to some other form of transportation.

Third, researchers believe Uber engages in dynamic price discrimination—that is, charging different prices to similarly situated customers—and manipulates consumer access to different tiers of service, so that some consumers see the less expensive UberPool service by default, while others see the more expensive UberX service, with the result that two different riders in the same location seeking transportation for the same route will be shown different pricing and different vehicles.[13]

Each of these examples disadvantages consumers, and each of them is possible because of the detailed personal data that Uber's platform collects.

This model of data usage in exchange for services is a vibrant one. Even when it has benefits, there can be hidden costs. For example, the financial service Credit Karma has generally received high marks from consumer advocates. A web- and app-based platform, Credit Karma users provide personal information like name, Social Security number, and date of birth, along with permission for Credit Karma to conduct the kind of inquiry that requires individual consent under the Fair Credit Reporting Act. In return, Credit Karma provides its users with a free copy of their credit score. Where Credit Karma earns its money, and how it maintains its popularity with consumers and lenders, is its matchmaking service, pairing up consumers who might be interested in new credit cards or other kinds of loans with lenders whose qualifications those individuals are likely to meet. The consumer doesn't get charged any fees by Credit Karma; instead, Credit Karma earns a commission from the lenders when a consumer signs up for a new credit card or takes out a new loan using Credit Karma's matchmaking. So far, so good.

In more recent years, however, Credit Karma and other tech platforms have begun offering free tax filing software. Where more established products like TurboTax charged users a fee, the terms of service for the newer "free" tax-preparation companies allow them to keep customers' personal data and use it for a variety of purposes. Tax returns are normally protected by law, but the information they contain —describing a person's mortgage payment, childcare expenses, charitable contributions, and the like—is valuable to data brokers to flesh out a more detailed profile of the individual's financial status, life circumstances, interests, and commitments. Users have an incentive to maintain their accounts: the free services often warn users that if they delete their accounts, they'll no longer have access to their tax returns or their underlying data. Since users might need this data in the future, to respond to IRS audits or for other reasons, the framework encourages them to maintain their accounts, which enables the platform to keep collecting more data. The upshot is that consumers are always trading something for these free services.[14] Perhaps ironically, TurboTax has often been criticized for its efforts to block the IRS from expanding its own, currently limited, free tax preparation software.[15] Now, it faces increasing competition from "free" services that have found other ways to monetize personal data. Either way, the individual bears the cost.

The thirst for data is limitless, it seems, and it isn't always commercial entities that are buying, selling, compiling, and sharing information. The Florida Department of Highway Safety and Motor Vehicles (FLHSMV) has reportedly been earning millions of dollars by selling Florida drivers' personal information to more than thirty private companies, including bill collectors, insurance companies, and data brokers. Florida residents have complained that, after their DMV data was sold, they were harassed by increased robocalls and other solicitations, from mailing and call lists that they hadn't been listed on before. So why would a government agency that, by nature, should be focused first on serving state residents, take actions that could compromise the privacy and peace of mind of its constituents? Apparently, the FLHSMV earned more than $77 million in fiscal 2017 alone through these data sales.[16] Perhaps the profit motive explains why, according to the DMV, there's no way for Florida residents who have driver licenses or state-issued ID cards to opt out of having that data sold.

One of the defenses offered by the FLHSMV was that it didn't sell data to marketing companies—but it does sell data to Acxiom. Acxiom, in turn, boasts that it has consumer profiles on over 700 million people around the globe, with more than 5,000 data elements from hundreds of sources combined into the "data products" that it sells to 2,500 companies worldwide.[17] Acxiom is only one of many companies in the data broker business, but as one of the biggest ones, its scale, scope, reach, and business model are worth taking a look at. Acxiom's website describes its mission as "identity resolution and people-based marketing."[18] The company promises that Acxiom's data can "power exceptional marketing experiences everywhere."[19] With buzzwords like "100 percent deterministic matching" and "multi-sourced insight," Acxiom claims that its products and services "form the 'power grid' for data, the critical foundation for people-based marketing that brands need to engage consumers across today's highly fragmented landscape of channels and devices."[20] In other words, since consumers' attention is divided, and their transactions are dispersed, across a multitude of platforms, apps, content-delivery systems, and more, Acxiom's mission is to collect the bits of data scattered across all of those places and package them into a holistic view that companies can use to reach consumers, grab their attention, and harness their purchasing power everywhere they go.

Acxiom's data profiles are used by companies in almost every major customer-facing industry, and it claims that these data profiles support "omnichannel, people-based marketing" that will enable companies to find the ideal audience for their products and services, individually tailor marketing messages to those people, measure the effectiveness of those marketing campaigns, and then convert those advertising campaigns into "successful data monetization" for Acxiom and its partners.[21] Acxiom's marketing materials explain that, to be effective in reaching individuals, companies need "to understand their offline and online presence, buying behavior, and interests."[22] Like an abusive spouse, Acxiom claims that this is the consumers' fault: "Consumers expect a connected experience."[23] According to Acxiom, it's these consumer expectations that drive companies' need for detailed individual behavioral profiles—not the companies' own desire for this data to carry out individualized marketing campaigns.

To meet these needs, Acxiom offers data packages for all kinds of interests and events. A typical example is this one from Acxiom's website: "Mother's Day—Families Spend Big on Mom. Don't miss out. Connect marketing to Department Store Dads or Generous Gentlemen to ensure the happiest Mother's Day for all."[24] In other words, it's all about looking for opportunities to generate enhanced interest in consumers who might be looking to spend for particular occasions, and then influencing how they approach that spending, their decisions about how much money to spend, and precisely what to spend it on. Of course, these practices aren't limited to any single company. The breadth and range of personal profiles offered by data giant Oracle is on display in its 2019 Data Directory, a 168-page catalogue of the different types of data collections that companies can buy.[25]

Although privacy advocates are hopeful that laws like Europe's GDPR will reshape the data-for-sale landscape and give individuals greater control of their data, the first EU enforcement case relating to this industry showed that it isn't yet clear how meaningful those new protections might be. In that 2019 case, the Polish data privacy regulator fined a digital marketing company €220,000 (about $250,000) for failing to properly inform some six million people that the company had gathered personal data about them. The striking feature of this fine: the marketer had scraped information that was already publicly available on websites with business registrations, national ID numbers, and legal events related to their businesses. In other words, it was fined for gathering publicly available data posted on government websites.[26]

Whether changes in privacy law will impact the data broker industry is unclear for now. Acxiom and Oracle are among the companies defending themselves against privacy complaints in three European countries.[27] The industry is also facing new transparency requirements in the United States, under laws passed in states like Vermont and California. The sheer size of this multibillion-dollar industry, however, suggests that it will be an uphill battle to persuade courts to impose stringent new privacy constraints.

CHAPTER 7
MINORITY REPORT

The Perils and Occasional Promise
of Predictive Algorithms

I n the 2002 movie *Minority Report*, Tom Cruise is on the run, under suspicion
as a future murderer based on the prediction made by a trio of semi-psychic
humans called "precognitives," whose ability to predict crimes before they happen has led to a revolution in policing.[1] In this future society, based on the precogs'
work, suspects are arrested *before* they commit crimes. The movie, like the 1956
science fiction story it's based on, raises questions about self-determination, personhood, autonomy, and free will. For many privacy scholars, as for *The Minority
Report*'s author, Philip K. Dick, privacy and self-determination are interwoven,
parts of the same fabric of individual life.

If the story were updated to 2020, the precogs would be replaced by
behavioral algorithms being run on a supercomputer somewhere—or, more
likely, on an instance of one of the commercial cloud services being offered by
Microsoft, Amazon, Google, and other such providers. The kinds of artificial
intelligence required to create and compute predictive algorithms would, not
too long ago, have only been possible on the hardware provided by high-end,
multibillion-dollar supercomputers like IBM's Watson. Today, however, the
capacity to create and run behavior-based algorithms has advanced so rapidly

that behavioral prediction algorithms are in widespread use for purposes so common, like targeted advertising, that they have become practically mundane.

In *Minority Report*, society has become a surveillance state. Everyday activities are recorded; spider-like robots carry out ID checks by pulling back people's eyelids and scanning their irises; every movement is tracked; and bespoke advertisements automatically populate digital billboards as prospective shoppers walk past them. All of this data is fed to the precogs, who engage in a shadowy, opaque divination process, and then spit out the results that change people's lives, perhaps preventing crime from time to time, and most certainly leading to the arrest and incarceration of people who, in fact, have not yet done anything wrong. Through the precogs' work, crime has been eradicated and society at large has accepted the role of widespread, continuous surveillance and behavioral prediction.

The algorithms of today threaten to do the same: to categorize us into narrow boxes in which our individual dignity is devalued, our self-determination is diminished, and, in extreme cases, our very liberty and autonomy are taken away from us.

Behavior-based algorithms are no longer simply being used to describe the past actions of individuals and groups. Now, these algorithms are forward-looking, predicting our likely next steps. And an increasing number of them are teaching themselves how to reach new conclusions. Behavior-based algorithms are being used in a variety of contexts. For example, a suicide text line uses algorithms to triage which calls are more serious and urgent. Medical researchers are using data from computer mouse tracking to predict whether a person will develop Parkinson's disease.[2] Parole boards are using algorithms as part of their decision-making process to predict which inmates are most likely to re-offend.[3] And in a growing number of cases, algorithms are reaching conclusions that are tainted by the same kinds of racial and gender bias that pervade society as a whole.[4]

US law has very little to say, at this point, about the use of behavior-based algorithms to make or support predictions and decisions of these sorts. European law, however, has begun to address this practice, but it isn't clear whether the regulators' approach will have a meaningful effect on the greatest risks posed by these programs. These risks include the ways that self-teaching algorithms, or "black boxes" (more on these below), reach decisions that pigeonhole our identities and deprive us of our autonomy, stripping us of the right to be left alone today and of the opportunity for self-determination tomorrow.

HOW ARTIFICIAL INTELLIGENCE AND MACHINE LEARNING WORK

The terms "artificial intelligence" and "machine learning" are sometimes used interchangeably, but there are important differences between them.[5] Artificial intelligence (AI) can be broadly thought of as the ability of computer software to make "smart" decisions and control devices in "smart" ways. A good example is the AI system in cars that adjusts your speed when you get too close to the car in front of you. The AI isn't sentient; your car isn't thinking for itself. Instead, the car has a set of sensors that detect your car closing the gap with another vehicle, and your car's AI signals the driver-assist technology to execute a set of preprogrammed responses: apply the brakes, activate a warning light, or turn off cruise control. The car is operating "smart" technology in the sense that it has detected conditions that meet a set of programmed parameters, and the car's telematics response frees the driver from having to be the sole set of "eyes and ears" to detect a potential hazard. But the car isn't "thinking"; it isn't exercising independent judgment, or acting in ways that the driver can't predict or understand. It's following a programmed routine.

Machine learning (ML), on the other hand, involves precisely the kind of self-teaching computer programs that we don't fully understand. At its core, ML is a subset of AI in which software programs are given data, and based on that information, the computers teach themselves. Much of ML is based on a computational field known as neural networks, computer systems that have been trained to classify information in the same way that a human brain does.[6] Like humans, neural networks take in information and make probabilistic judgments about it. As they make those judgments, they learn from a feedback loop that tells them—either based on more data, or on input from a human—whether their probabilistic judgment was correct. Over the course of many iterations, the ML program "learns."

In many cases, the judgments being made by these AI/ML systems are benign. They read a piece of text and gauge the mood of the person who wrote it, whether the words are intended in an angry, happy, or sarcastic tone. They "listen" to a piece of music and assess whether it's likely to make a human listener happy or sad, and they recommend music with similar characteristics to create streaming radio stations with a coherent "vibe."[7]

Machine learning programs have often been described as "black boxes," because even the system's own designers can't explain precisely why the computer reached a particular conclusion.[8] The designers can explain what the purpose of the particular neural network is; they can describe the information used to train the algorithm; they can describe other outputs from the computer in the past. But—just as we often don't know *why* humans reach the conclusions that they do—computer scientists frequently can't explain *why* an ML system reached the conclusion that it did.

Sometimes, those results are troubling, indeed.[9] When an algorithm botches the auto-suggest feature on a music playlist, the worst consequence is minor annoyance. But algorithms are increasingly being used in settings that have the potential for two kinds of harms: to invade privacy and to negatively, and baselessly, impact individuals' lives. For example, when algorithms have been used by parole boards to predict the likelihood of recidivism, the predictions have been less accurate than a purely human assessment.[10] When Amazon implemented algorithms to screen job applicants, they have heavily favored white male applicants, perhaps because the algorithms were trained on a data set in which the majority of hires historically were white men.[11] When the state of Arkansas began using AI to assess the number of caregiver hours Medicaid patients should be entitled to, the results from a proprietary algorithm were a drastically reduced level of care.[12] When the city of Houston began using a third-party algorithm on student data as a means of making decisions about teacher evaluations, it turned out that not a single employee in the school district could explain, or even replicate, the determinations that had been made by the algorithm.[13] AI in facial recognition systems is notoriously bad at identifying non-white, non-male faces, and countless AI sets have been trained on data that, because it is reflective of historical biases, has a tendency to incorporate and perpetuate those very biases that Western democracies are trying to break free from.[14] And in an oddly poignant twist, when Microsoft released an AI machine learning bot, named Tay.ai, onto the internet to test its ability to interact in human-like ways, the bot had to be taken offline within less than twenty-four hours. The bot had been so trained by the cesspool of online bad human behavior that it was spewing racist and hate-filled ideologies.[15]

DESPITE THE RISKS, AI ISN'T ALL BAD

DoSomething.org is a nonprofit organization that regularly sent out texts to young people who were interested in its messages encouraging action to bring about social change.[16] In 2011, one of DoSomething's communications managers received a text, out of the blue, from someone on their mailing list. The person was in crisis: they reported they had been repeatedly raped by their father, and they were afraid to call the nation's leading hotline for sexual abuse. The DoSomething employee who received the text messages showed them to Nancy Lubin, DoSomething's CEO. And Lubin knew she needed to *do* something.

Within two years, Lubin had launched the nation's first text-only hotline, the Crisis Text Line (CTL).[17] Within four months of its launch, it was providing services for individuals in crisis across all 295 telephone area codes in the United States, and by 2015, the 24/7 service was receiving 15,000 text messages per day. Many of the texts are about situations that, while painful and difficult, do not indicate the texter is in any immediate danger. But about once a day, someone sends in a text indicating that they are seriously and imminently considering suicide; this is someone who needs active, immediate intervention. One of the reasons the text line has gained so much traction is that, according to research, people are more willing to disclose sensitive personal information via text than in person or over the phone.[18]

CTL's second hire was a data scientist who approached the challenge of effective crisis message triage first by talking with volunteer crisis counselors around the country to learn from their perspective, and then collecting data and developing algorithms to generate insights from it. According to a lengthy review of CTL that was published in the *New Yorker* in 2015:

> The organization's quantified approach, based on five million texts, has already produced a unique collection of mental-health data. CTL has found that depression peaks at 8 P.M., anxiety at 11 P.M., self-harm at 4 A.M., and substance abuse at 5 A.M.[19]

The sheer volume of information available to CTL is striking in the mental health field. By way of contrast, the American Psychiatric Association's journal, *Psychiatric News*, published an op-ed in 2017 calling for mental health research to carry out more big data analysis as a way to understand trends, spot patients who might be at risk, and improve delivery of care. According to the article:

> When it comes to big data science in mental health care, we live in the dark ages. We have only primitive ways to identify and measure mental health care, missing out on opportunities to mine and learn from the data using strategies that can create new discoveries we have not yet imagined.[20]

The author argued that ethical data collection would benefit individual patients as well as overall cases, and that the primary health-care system was positioned particularly well to help facilitate those goals.

As it happens, the Crisis Text Line might be paving the way for precisely those innovations. In 2015, CTL was looking into developing predictive algorithms. By 2017, CTL had succeeded in doing just that, with some surprising results. For example, after analyzing its database of 22 million text messages, CTL discovered that the word "ibuprofen" was sixteen times more likely to predict that the person texting would need emergency services than the word "suicide."[21] A crying-face emoji was indicative of high risk, as were some nine thousand other words or word combinations that CTL's volunteers crisis counselors could now be on the lookout for when interacting with the people texting in. CTL's algorithmic work didn't stop there. An article published in 2019 noted that CTL had analyzed 75 million text messages, and from its analysis had generated meaningful data about the most effective language to use in a suicide intervention conversation. Based on those findings, CTL issued updated guidance to its counselors, telling them that it was helpful to express or affirm their concern for the person but that incorporating an apologetic tone ("I hope you don't mind if I ask, but . . .") was less effective.[22]

By 2017, Facebook announced that it, too, was going to use artificial intelligence in order to assess its users' emotional states.[23] Whatever the merits of the intention behind this move, the privacy implications of Facebook's decisions were very different from the Crisis Text Line. The CTL had both a privacy model and a practical context that protected its users' expectations and needs. For example, a person who sent a text to CTL received a reply with a link to the privacy policy and a reminder that they could cut short the conversation at any time by typing "stop." The person reaching out to the CTL was already in some sort of practical difficulty or emotional distress, and they had proactively reached out to CTL. In other words, they knew exactly who they were texting and why and could anticipate that their communications would be reviewed specifically with an eye toward trying to understand what kind of help they needed, and how quickly they needed it. And CTL's review of the content of text messages served that purpose only: to provide help. This is a very different purpose than serving up targeted advertising based on the content of the users' messages—which is the core business purpose for so many algorithms that run across digital platforms. CTL's focus on supporting positive mental health outcomes was evident in other aspects of the organization's structure as well, from its nonprofit status to the fact that it has a Chief Medical Officer, a Clinical Advisory Board, and a Data, Ethics, and Research Advisory Board, all composed of experts in the fields of medicine, psychiatry, social work, data science, biomedical ethics, and related fields.[24]

Facebook's foray into mental health assessments began in 2016, when it announced it was adding tools that would let a user flag messages from friends who they believed might be at risk of suicide or self-harm, teeing up the post for review by a team of Facebook employees.[25] According to the *New*

York Times, these Facebook tools marked "the biggest step by a major technology company to incorporate suicide prevention tools into its platform."[26]

The Facebook model, although initially well received by the press, was almost the antithesis of CTL's approach. Facebook's worldwide userbase was using its platform for other purposes: to share pictures of their travels, keep in touch with their friends, advertise their business, and find news and laugh at memes. Before these new features were rolled out, Facebook's users had no reason to expect that algorithms would run across their posts, likes, and shares with an eye toward assessing their mood or mental health. And they certainly weren't looking to Facebook to use algorithmic conclusions to serve up tailored advertising based on the AI's assessment of what they might need. The circumstances were ripe for misuse: it wasn't hard to imagine people making posts, and flagging them, as pranks. And it isn't a far stretch to imagine Facebook serving up ads to users when their mental state leaves them at their most vulnerable, enticing them to make impulse buys.

The hazards of Facebook's mental health activities were underscored by the fact that the company had been penalized as far back as 2011 for using personal data in unfair and deceptive ways, and had more recently tested whether it could manipulate its users' emotions—make them feel more optimistic or pessimistic—by changing the news stories that popped up in their feeds.[27] When confronted with an outcry over those past abuses, Facebook announced it would provide its developers with research ethics training and that future experimental research would be reviewed by a team of in-house officials at Facebook—but that there would be no external review or external advisory body, nor any disclosure of the decision-making relating to Facebook's use of its platform to carry out human psychological or behavioral research.[28]

When the company made its 2017 announcement that it was using artificial intelligence to assess whether its users were suicidal, it noted that the tools weren't being applied to users in the European Union, as this kind of behavioral profiling would likely have been impermissible under European data privacy law.[29] Although Facebook presented this effort as a way to provide a socially beneficial service to its global user base, the company did not provide details on how the AI was tested or validated, or what privacy protections were in place. It did, however, note that, in geographic areas where the tool was deployed, users didn't have the ability to opt out.[30] Further, Facebook wasn't planning to share any results with academics or researchers who might be able to use the information to broaden understanding in the suicide prevention and crisis intervention fields.[31]

On the contrary, there were reports in 2017 that Facebook was showing advertisers how Facebook could help them identify and take advantage of Facebook users' emotional states. Ads could be targeted to teenagers, for example, at times when the platform's AI showed they were feeling "insecure," "worthless," and like they "need a confidence boost," according to leaked documents based on research quietly conducted by the social network.[32] Despite the lack of research ethics of privacy protections, and the apparent profit motive for this move, one survey showed that, by a margin of 56 to 44 percent, people didn't view Facebook's suicide-risk detection to be an invasion of privacy.[33] One would have to expect that, if the same poll were done today, the results might be very different, as Facebook spent much of 2018 and 2019 fending off a series of damaging news reports and investigations by legislators and regulatory bodies around the world relating to concerns that its privacy policies were lax at best—and unconscionable and illegal at worst.

Back to the good news: the Crisis Text Line seems to offer an example of the ways in which large sets of data can be collected and analyzed for insights relating to highly personal, sensitive topics, and make valuable contributions to enhancing health and well-being, without compromising the privacy of the individuals whose data is being reviewed. As leading voices in the field of AI continue to reiterate, the goal shouldn't be to prevent altogether the development and use of AI. Rather, the goal should be for humans to follow a very ML-like process of learning from the feedback provided both by failed AI experiments and by their successes, and use those results to continuously improve the approach. There's promising work underway from academic researchers and the private sector on two fronts: how to understand what happens in the black box, and what kind of AI code of ethics is needed in the design and use of machine learning systems. The EU has legislated protections against automated decision-making, and the US Congress has been holding hearings on AI, ranging from concerns over personal privacy and bias in AI to how the technology can be effectively used. Perhaps, then, if we continue this feedback and improvement loop, it will prove that we are as capable as our machines.

CHAPTER 8
DIFFERENTIATING THE REAL FROM THE FALSE

Social media platforms have become the front lines of viewpoint manipulation and propaganda. Media-conscious influencers can hire "black PR" firms to develop content, create fake follower accounts, and manipulate social media engagement in order to create carefully curated public images for politicians, entertainers, or anyone else willing to pay for their services.[1] Savvy individuals, campaigns, and governments are using the widespread reach of social media and the built-in thought bubbles created by our decisions about whom to friend and whom to follow as a means of leveraging our personal data to change the ways we think about the world.

To be clear: this kind of influence operation is different from traditional news outlets that post perspectives, information, and sometimes even lies. Traditional content publishers either "push," or broadcast, their information to receptive users, or let users "pull" content by visiting the publishers' websites, clicking on links, and the like. Those publishers and sites rely on a marketplace of ideas: they provide information, and leave it up to their readers, viewers, or listeners to form an opinion about it. When it comes to social media manipulation and microtargeting, however, platforms, advertisers, and public relations firms use an array of tools to subtly influence how we receive the content that is being pushed to us. They use detailed, individualized profile information to identify whom to target—based on our demographics, shopping habits, online friends, and more—and then assess how to *most effectively* target us, based on the platforms' assessments of our individual personality traits, such as how much we sympathize with people who are different from us and how drawn we are to autocratic tendencies.

This kind of individually targeted viewpoint manipulation has been used to shape public opinion about the protests in Ferguson, Missouri, and Baltimore, Maryland; to influence Britain's Brexit vote and the 2016 US presidential election; and to try to sway the elections in Germany and France in 2017 and 2018. These influence campaigns rely on a mix of tools: automated bots and human trolls create propaganda and fake news; they leverage users' personal data to identify specific individuals likely to be susceptible to their messaging; and, at very little cost, they flood users' feeds with a nearly endless supply of posts that make it appear as though there's a groundswell of opinion in favor of a certain candidate or idea.

Social media platforms are making it possible for manipulators to use our own personal data as the most formidable weapon against us, and to great effect. National security experts fully expect that foreign adversaries will try to interfere with the 2020 US election cycle. The risk isn't limited to foreign governments. In 2019, Facebook's founder, Mark Zuckerberg, claimed that an Elizabeth Warren presidency—with a renewed focus on antitrust regulation, privacy protection, and wealth taxation—would be the social media platform's worst nightmare.[2] Zuckerberg's comments raise a whole new set of concerns about the ways in which the owners of these social media empires are not only profiting from others' use of the platforms to manipulate opinion, but also how the platforms themselves might target us with individually tailored messages designed to shape our opinions to suit their own political goals.

Social media platforms are making it possible for manipulators to use our own personal data as the most formidable weapon against us, and to great effect.

In March 2019, Robert Mueller's Office of Special Counsel released a two-volume report on Russia's interference with the 2016 US presidential election.[3] Based on nearly two years of investigation, Mueller and his team concluded that the Russian government interfered in a "sweeping and systematic fashion." One prong of that interference was an "active measures" campaign that included using fake social media accounts to sway opinion in the United States. The Russian government worked through a St. Petersburg–based troll farm known as the Internet Research Agency (IRA), run by a close associate of Russian president Vladimir Putin. Beginning in 2014, IRA employees created fake accounts on Facebook, Instagram, Twitter, and other social media platforms. The accounts, which varied widely, were designed to look like authentic accounts associated with real people living in the United States.

One of the most widely followed IRA trolls was a Twitter account with the handle @TEN_GOP—an account that fooled hundreds of thousands of Twitter users into believing that it was the "unofficial account" of the Tennessee Republican party. Another account, @Jenn_Abrams, famously duped her 70,000 Twitter followers. The fake account posted tweets about pop culture, ballistic missiles, the Confederate flag, and Rachel Dolezal, a white woman who's known for self-identifying as black. "Jenn Abrams's" persona was so persuasive that her tweets were featured in articles in *USA Today*, the *New York Times*, *The Daily Caller*, *BuzzFeed*, and a host of other US and international news outlets. @Jenn_Abrams followed a modus operandi common to many of these troll accounts: build up a following with an entertaining and engaging online persona that posted content about nonpartisan issues—celebrity gossip, general news—and then, after attracting a substantial following, start pushing out deeply divisive content on wedge issues in American politics, like immigration, race, gay rights, and, closer to the 2016 election, content deeply critical of, or spreading conspiracy theories about, Democratic nominee Hillary Clinton. Abrams was particularly successful at creating a total package of a person; in addition to her Twitter account, she also had a personal website, a Medium page, a Gmail address, and a GoFundMe page.[4]

In 2019, the *Columbia Journalism Review* published research showing that most major news outlets had at some point in time unknowingly quoted

Russian hoax accounts, usually in stories reporting on public reaction to recent events. News outlets with a more left- or right-leaning bent, such as *The Daily Caller* and *Huffington Post*, meanwhile, were more likely to report on the often incendiary social media posts. But the challenge of differentiating authentic accounts from fake ones—the problem of differentiating the true from the false—hit major mainstream news outlets in the center of the political spectrum as well, including such highly regarded ones as the *New York Times*, NPR, and the *Washington Post*.[5]

The Russian active measures campaign was carried out on every major social media platform. On Facebook, for example, the IRA controlled 470 accounts that made over 80,000 posts between 2015 and 2017, reaching some 126 million people.[6] These accounts began creating and sharing openly pro-Trump and anti-Clinton posts and reaching out to pro-Trump groups. They even contacted Facebook followers via private direct messages to encourage them to show up for live, in-person rallies in the United States—rallies that were staged by the IRA, acting through its Russian-government-backed trolls from St. Petersburg, thousands of miles away.[7]

Neither the criminal indictments against the IRA nor the widespread news coverage about Russia's fake social media profiles and influence campaign seem to have had any deterrent effect. Russia's troll farms were accused of attempting to influence the 2017 French presidential election and the 2017 German federal election, and of stoking France's "yellow vest" protests in 2018. Russian troll farms were also implicated in using Facebook, Twitter, and YouTube accounts in their attempts to sway European Union elections in 2019.[8] The playbook was the same as it had been in the United States in 2016: suppress voter turnout, deepen political divides, and advance a far-right policy agenda through fake accounts, disinformation, and the exploitation of local political divisions that already existed. Adding insult to injury, in 2019, a Russian troll farm that was a close cousin to the IRA, the Federal Agency of News, filed a lawsuit in US federal court, charging Facebook with violating its First Amendment rights for kicking it off the platform.[9]

The problem is increasingly well documented. Reports from the UK Parliament have laid out in excruciating detail the ways in which social media disinformation is fueling social discord and political divisiveness, as well as uninformed or misinformed political thinking. The Australian Competition and Consumer Commission has issued a report describing the ways in which social media platforms are not only spreading misinformation but also driving legitimate news outlets out of business.

It's important to be clear what the risk is here: the pernicious fact of social media's role in this shifting information landscape is that social media makes it so easy to feed the trolls. Because social media platforms know so much about our individual behavior, interests, personality traits, and inclinations, and because of the way that their advertising and engagement models work, they are the perfect mechanism for people with a message—including distorted, damaging, false, or divisive messages—to individually target their content toward each one of us.

Meanwhile, the platforms are eschewing any responsibility for their role in fomenting social divisions or political unrest. Facebook has mounted an aggressive defense to the class action lawsuits filed against it by users who allege that the platform has violated their privacy by providing their personal data to the political advertising firm Cambridge Analytica, which then used it for microtargeted messaging without the users' knowledge or consent.[10] As described by the court overseeing the litigation, "Facebook argues that people have no legitimate privacy interest in any information they make available to their friends on social media." Further, according to the plaintiffs' allegations, although Facebook had a nominal policy restricting third-party access to Facebook user data, in effect no such policy existed at all, because "with the tens of thousands of app developers who interacted with users on the Facebook platform . . . Facebook was intent solely on generating revenue from the access it was providing."[11]

It's hard to know exactly how much money Facebook earns from political ads compared with other kinds of advertising, but when confronted with concerns, Mark Zuckerberg has consistently taken the position that, unlike some other platforms, Facebook will continue to sell advertising to political campaigns.[12] It will also continue to benefit from the indirect revenue generated by increased user engagement and time-on-platform that's created by unpaid troll accounts. And in a twist that underscores the dangers that Zuckerberg's personal views could be used to drive Facebook's content moderation, it turns out that some of the biggest political advertising dollars spent on Facebook are spent by Zuckerberg himself.[13]

HOPE ON THE HORIZON?

Think tanks like the Alliance for Securing Democracy at the German Marshall Fund have pointed to the destabilizing effect across all of the Western democracies of authoritarian and nationalist movements that are being fueled in part by information operations that include these individually directed and microtargeted political messaging campaigns.[14]

Academics like Kathleen Hall Jamieson at the University of Pennsylvania and Briony Swire-Thompson at Northeastern University have done extensive research on the phenomena of fake news, Russian troll farms, and targeted social media political advertising. Jamieson, a prominent expert in public policy and political communications strategies, was one of the first to do a comprehensive social science analysis of the impact that Russia's information operations had on the 2016 US elections. Her conclusion: the Russian active measures campaign was powerful and effective in shaping US public opinion, and likely changed the outcome of the election in favor of Donald Trump.[15] Swire-Thompson's research has a somewhat different focus, examining the cognitive mechanisms behind our susceptibility to "fake news." In one of her recent papers, she notes that when people are attempting to assess the trustworthiness of information, we too often default to cognitive biases like, "They might be a liar, but they're my liar."[16]

In looking for solutions, some policymakers have suggested new legislation like the Honest Ads Act, which would require political advertising to carry funding disclosure statements much like those that are required for radio or television ads. Others have correctly pointed out that, while microtargeting happens through paid advertising, there should also be more content moderation, flagging bots and trolls, taking down inauthentic accounts, and removing offensive content.[17] Other policymakers focus on the need to emphasize critical thinking skills in school but tend to overlook the research showing that younger people are generally more savvy about the need to be skeptical of internet content. It's older voters who are more likely to be swayed by slanted content or deceived by outright falsehoods that are posted online.[18] Given those demographic distinctions, it's clear that K–12 or college education alone isn't nearly enough to combat a cultural receptivity to misinformation, especially when it has been microtargeted to individually exploit us.

Finland might have an answer. The country's population of 5.5 million has been wary of Russian threats to its national security ever since declaring its independence from Russia over a century ago.[19] As part of an initiative launched in 2014 to counter Russian information warfare that attempted to stoke divisions within Finland over issues such as immigration, the European Union, and NATO, Finland has taken a multi-pronged approach to citizen education that includes classroom projects at all age levels as well as adult education programs in local community centers. The lessons address how to identify bots on Twitter, how to spot deep fake videos, how to spot slanted news coverage and identify an outlet's or an article's biases, and more. These measures alone may not be enough to counter the ways in which our individual information is being used to manipulate our views. But it appears to be having positive effect, both in maintaining civil discourse in Finland and in preserving free and fair elections there. Countries across Europe and from other parts of the world are looking to Finland's experience to create critical thinking and public education programs of their own. As part of a larger toolkit, Finland's lessons could contribute in important ways to countering the negative effects of platforms that know the details about us entirely too well.

SECTION III

Power Play: How Personal Data Exacerbates the Power Imbalances in Everyday Life

The first few months of 2017 brought significant news stories in how big data and technology are shaping the lives of employees around the globe. On the same day in January, two headlines offered a striking contrast. In one, a California woman sued her boss for insisting that she install a mobile phone app that tracked her location outside of work hours. In the second, a new law in France gave employees the right to bargain for limits on their out-of-hours availability, creating at least the possibility of a right to ignore email after certain times. Around the same time, the US House of Representatives voted that workplaces could require staff to undergo a genetic test in exchange for lower health insurance premiums. And reports surfaced that 25 percent of employers are now using predictive analytics in hiring decisions. The possible uses of data and devices are expanding faster than employment laws and collective bargaining can keep up.

The power imbalances exacerbated by data tracking and analysis aren't limited to the workplace. Schools are using facial recognition technology to measure student engagement, colleges are turning to predictive algorithms in making admissions decisions, and personal digital assistants—that also serve as private surveillance devices—are proliferating in classrooms where preschoolers have little opportunity to provide meaningful consent. Readily available "spyware" has escalated the dangers of intimate partner violence and reinforced the ways in which governments like Saudi Arabia allow male guardians to exert chokehold control over the women in their custody. These power imbalances aren't new, but today's data-driven technologies are rapidly widening the divide, consolidating power in the hands of those who already have it, and further diminishing the sense of autonomy, independence, and personhood of those whose digital data is being turned against them.

CHAPTER 9
IT'S 11 PM. DO YOU KNOW WHERE YOUR EMPLOYEES ARE?

I n today's data-driven economy, employers have access to more personal information about their workers than ever before. The uses, and potential misuses, of this information arise in a variety of ways. Bosses use mobile phone apps to keep tabs on their employees' whereabouts after hours. Human resources departments and workplace managers have access to sensitive personal health information as part of employee wellness programs. Companies are carrying out increasingly granular recordings of their employees' conversations, locations, and movements in the workplace, even offering staff the ability to get company-provided RFID chips embedded underneath their skin.

LOCATION TRACKING

The trend toward tracking employee location data began in an analog era, with transportation safety laws requiring long-distance truckers to keep logbooks of their mileage, with the aim of limiting the number of hours drivers would be on the road without sleep. As any long-haul trucker can tell you, there were often two sets of books: the real ones shown to the employer to show fuel costs and jobs completed, and a second set shown to federal regulators, which sometimes displayed considerably lower mileage numbers. In the days before LoJack, keeping two sets of handwritten paper logs was a viable way for truckers to maximize their distance—and pay—when they weren't in strict compliance with the safety rules set by Uncle Sam. In the 1960s, when artists like Dave Dudley were making truck-driving songs with titles like "Speed Traps, Weigh Stations, and Detour Signs," or "Six Days on the Road," long-haul truckers most likely couldn't have imagined the extent to which millions of people in hundreds of lines of work would one day have their locations tracked by their employers on a minute-to-minute basis.

In the modern economy, workplaces are demanding an entirely new level of data-intensive monitoring for their employees. Staff are sometimes required to install data tracking apps on their personal phones, use employer-provided phones that track their locations during their after-hours time, and use high-precision fitness wearables that track their movements.

Perhaps because of the inherent power imbalance driven in part by decreasing levels of collective bargaining and a global economy in which many employers are able to insist on take-it-or-leave-it conditions, there's been surprisingly little pushback on these intrusive practices so far. Myrna Arias, however, was one notable exception. Arias had been recruited from her previous company, NetSpend, by John Stubits, Regional Vice President of Sales at money transfer company Intermex. Stubits persuaded Arias to take a sales executive position at Intermex in Bakersfield, California, where she worked for three months in 2014. According to the Complaint[1] Arias filed in Bakersfield County Superior Court, Arias met all of her sales quotas and did an excellent job in her new role.

In the Complaint, Arias states that, in April 2014, Intermex required all of its employees to download and install a mobile workforce management app called Xora. Among other features, Xora used GPS-enabled location tracking. Arias and her coworkers asked whether the company would use the app to monitor their off-duty whereabouts. According to the Complaint,

"Stubits admitted that employees would be monitored while off duty and bragged that he knew how fast she was driving at specific moments ever since she had installed the app on her phone."[2] Arias explained that she had no problem with Intermex monitoring her location during working hours, but "she objected to the monitoring of her location during off hours and told Stubits this was an invasion of her privacy. She likened the app to a prisoner's ankle bracelet and informed Stubits that his actions were illegal."[3] According to the court filing, Stubits said she shouldn't complain because she was earning more money at this new job, and she needed to keep her phone on 24/7 in order to answer client calls. That same month, Arias uninstalled the app from her phone, and just a few weeks later, she was fired.

Arias's lawsuit included a number of claims, including a count for "Invasion of Privacy—Intrusion into Private Affairs," relating to Intermex's off-duty GPS tracking. The heart of Arias's complaint was that information about her whereabouts and conduct while off-duty were "private and highly confidential," that "a reasonable person would have an interest in maintaining the confidentiality of such information," and that her employer's intentional invasion of that privacy would be "highly offensive to a reasonable person."[4]

The lawsuit, demanding over $500,000 in damages, settled out of court in November 2015.[5] But it left in its wake a number of questions that are still live and pressing today, all of which center around a core theme: When, and to what extent, does the law currently allow employers to track their employees' activities and locations? When, if ever, *should* that kind of off-duty tracking be allowed? What, if anything, are courts and legislatures doing to protect the employees' privacy? In a country with fewer unionized jobs and labor laws that vary widely from state to state, many employees have little recourse if an overbearing employer demands access to their location information at all times, even when off-duty and using non-work vehicles or phones.

When the case was first filed, some commentators focused on the growing trend toward companies expecting employees to be on call and available twenty-four hours per day. That almost seems a quaint notion now, as the

24/7 availability expectation has become nearly ubiquitous across government, for-profit, and nonprofit employers alike. Today's data-driven technologies provide countless ways to facilitate those demands and to support the expectation that we'll respond to work demands at any time of day or night. If the GPS on our work-issued phone is constantly showing where we are, it's all the easier for our employers to demand that surely we should be able to respond to a call, text, or email right this very minute.

It's no secret that studies are consistently showing the "toxic" effects of constant demands to be on call, available, plugged in, and with devices turned on at all hours of the day.[6] In response to these concerns, France passed a law that took effect on January 1, 2017, establishing a "right to disconnect."[7] Under the legislation, workers at establishments with more than fifty employees have the right to bargain over the terms of their after-hours availability.

When and whether an employer can legally, or ethically, monitor our behavior depends on a host of variables, each of which can be mixed and matched like a logic game. The outcome varies according to whether the monitoring occurs in personally owned or company-owned vehicles, during on-duty hours or off-duty hours, via personally owned or employer-issued mobile phones, and more.

It isn't just off-duty concerns; during on-site shifts, workplace surveillance can reinforce power imbalances, exploiting workers and invading their privacy at the same time. Amazon's warehouse practices provide a case in point. For years, Amazon employees have complained about working conditions.[8] Faced with ever-increasing order fulfillment requirements, warehouse staff who have to process a new package every nine seconds say they're peeing in bottles and skipping breaks in order to meet their quotas.[9] There are large-screen broadcast photos of employees who were caught stealing, their images stamped with the words "Arrested" or "Terminated."[10] What does this have to do with privacy? Employees are searched upon entering and exiting the facilities, stationary surveillance cameras track their every move, and some employees are required to wear closed-circuit body cameras, much like the body cams used by police forces nationwide.[11] When an Amazon employee died from a heart attack on the warehouse floor in 2019, his brother asked, "How can you not see a six-foot, three-inch man lying on the ground and not help him within 20 minutes? A couple of days before, he put the wrong product in the wrong bin and within two minutes management saw it on camera and came down to talk to him about it."[12] By all accounts, privacy-intrusive technology is robustly employed to prevent theft and to force greater productivity, but not to achieve improvements in worker health, safety, well-being, or morale.

While Amazon insists it offers a positive work environment, its patent applications tell a different story. In 2018, Amazon registered patents for workplace bracelets that would use electronic signals to register each piece of inventory touched by an employee and vibrate when the device needed to signal the worker. Relying on hypersonic pitches, the electronic signal from the bracelet would interact with modules placed in locations like inventory bins. In theory, this could free up workers' hands because they wouldn't have to carry inventory scanners, and the bracelet could vibrate an immediate notification if an item went into the wrong bin. But it could also be used to provide even more granular data about the speed of each employee's work, their precise location on the factory or break room floor, and how long they spent in the toilet with inventory modules.[13] Workplace advocacy groups have pointed out the risks: increased anxiety and stress in an already high-pressure environment, lower morale, potential discrimination based on physical disabilities or limitations, and increased challenges in attracting and retaining employees. Their analysis just underscores the Amazon workers' lived experience. According to one account of Amazon warehouse worker training back in 2013, "My initial thought was this is prison, the comparisons were obvious. I felt like asking anyone sitting by me or standing in line next to me 'so, what are you in for?' It would have been a completely appropriate question."[14]

Wearables and mobile phone apps are more than intrusive enough. But a handful of companies have taken the next step. In 2017, Epicenter, a Swedish company, began implanting RFID chips under the skin of any employees who volunteered to have this done. Epicenter encouraged use of implanted RFID chips that workers could use to "unlock doors, operate printers, open storage lockers and even buy smoothies."[15] By 2018, a company in Wisconsin had followed suit, with fifty workers signing up to have RFID chips embedded under their skin.[16] It raises a host of questions: Can the chips (or the wireless network their signals ride on) be hacked in ways that create false records about the chip's bearer? Can companies install surreptitious RFID readers in places the employee is unaware of, tracking the worker over a wider geographic area? What happens to the embedded chip when the employee resigns or moves on to another job? However, none of these concerns seem to be top considerations. The response of many analysts when asked whether more companies in the United States might adopt such approaches? Not likely—because it isn't necessary. There's already so much smart-badge tech available that embedding chips doesn't really add value. For the past half-decade, smart badges have been able to track where you go and how much you fidget; some even have microphones to register the emotions in your voice and how often you talk.[17]

The inescapable result: corporate executives on the hunt for higher profits are bulldozing past any quaint ideas of employee privacy. Companies are accelerating their use of surveillance technology, hoping to drive increased productivity and lower costs. The encroachment on workers' sense of personhood and autonomy may yield near-term economic gains for the company but at a long-term cost to society.

THE PRIVACY IMPLICATIONS OF WORKPLACE WELLNESS

"Workplace wellness" programs have grown in popularity since the passage of the Affordable Care Act. By 2019, workplace wellness had become an $8 billion industry, with programs offered at 84 percent of large organizations (two hundred or more employees) and 50 percent of smaller ones.[18] Among employers who offered health benefits, 11 percent collected fitness data from workers through mobile apps and devices such as Apple watches, Fitbits, and other fitness wearables.[19] Often, these health-promotion programs are coupled with financial incentives, which can include gift cards, merchandise (like fitness trackers), cash, contributions to health-related savings accounts, or discounts on health insurance.[20]

A EUROPEAN VIEW ON COMPETING INTERESTS: PRIVACY AND PAY

At a 2019 conference on data privacy laws in the United States and Europe, lawyers and regulators were giving their perspectives on the use of cameras and other kinds of surveillance in the workplace. The US panelist argued that cameras were essential: "When you're talking about equipment on a factory floor, this is expensive equipment and we need to be able to understand what may have caused something to break or how something went wrong. Or maybe an employee sues, alleging some kind of misbehavior. So sometimes you need to be able to review that footage." She had delivered a persuasive, and at times impassioned, articulation of all of the reasons why it was necessary to be able to document what happens in the workplace. The crowd in attendance nodded in assent.

Then the German representative spoke up: "When someone asks me about whether they can install cameras in the workplace, I often ask the bosses, 'Has anyone ever become more productive or happier at work because they're being watched by a camera?'" There was an audible pause just before the audience burst into laughter. The answer was obvious: Of course not. She continued, "I also ask the bosses, 'Would you be happier or more productive at work if there was a camera in your office?'" Again, the answer was obvious.

The German panelist didn't argue that workplace cameras should never be used, only that they shouldn't be a first recourse or a default option. Instead, she contended, each employer should give careful thought as to *why* they wanted to install this equipment. In the same matter-of-fact way she had delivered her other comments, she noted that, "My impression is that people want to sell the cameras, so they go to the businesses and tell them they should use the cameras. But the businesses haven't really thought through what they're going to do with that data, or what the point of it is. This lack of thought is a problem."

The Irish and Belgian panelists described a number of relatively straightforward mitigations that are largely absent from the US approach to workplace monitoring. The Irish delegate noted that she had very little concern about workplace surveillance video being retained for twenty-four hours; if surveillance was justified in the first place, then keeping the tapes for twenty-four hours was a short time frame that didn't represent any significant intrusion on privacy. But, if one takes the approach—as European law does—that it's necessary to balance employer interests against individual rights, then keeping those tapes for months on end, or indefinitely, would almost never be justified.

Despite these commonsense measures, European law isn't fully resolved on these issues. On the one hand, the European Data Protection Board (EDPB) has issued guidance on the process for analyzing decisions about when and whether to use camera surveillance in the workplace, pointing to the need for caution and balance when using video cameras to monitor employees at work.[21] On the other hand, a 2019 court case in Spain demonstrated how employees' other interests sometimes compete with their privacy concerns. In that case, a Spanish labor union had complained that employees weren't being awarded all of the overtime pay that was due to them. The European high court had previously ruled that employers must track the precise time their employees clock in and out of work, to ensure that workers are being paid for all of their time and that they're being granted the periodic rest breaks they're entitled to under law. The court didn't mandate any specific approach to measuring employees' time at work but did point out the inherent power imbalance between workers and employers as one reason why the methods used must be "objective, reliable, and accessible."[22] Video surveillance and biometric information can be effective ways to do that.[23]

These separate cases in Spain and Europe's highest court point to the reality of competing interests. Where overly intrusive surveillance can undermine workers' privacy rights, those very same technologies can be important means for protecting workers' wage and working condition rights.[24]

The American Association of Retired Persons (AARP) filed suit against the Equal Employment Opportunity Commission (EEOC) over the size of these wellness program incentives, arguing that the EEOC's rules for the programs essentially allowed employers to penalize workers who chose not to participate. The AARP won, causing the EEOC to revise the rules to cap incentives at lower limits that would have a less coercive effect. [25] Even following the litigation, however, the rules on incentives remain unclear and financial incentives can be substantial: as much as a 30 percent savings of health insurance premiums, which could amount to nearly $1,500 per year.[26] The size of the incentives has led privacy advocates, disability advocates, senior citizens' groups, and others to challenge the fundamental fairness of these programs. For privacy advocates, disproportionate financial incentives call into question whether providing the information demanded by these workplace wellness programs is truly "voluntary." For disability rights advocates, there are questions about whether the programs discriminate against individuals with physical limitations that prevent them from participating in some aspects of these initiatives, such as challenges to walk ten thousand steps each day, go to the gym three times a week, lose weight, or lower their blood pressure.

It's worth noting that, under these workplace wellness programs, the health-related information that employees provide is often unprotected by any laws. If the wellness program is offered through an insurance company, the protections over HIPAA will likely apply, because the insurance companies are "covered entities" that are bound to comply with HIPAA. But many programs are run by third-party vendors. These vendors operate outside of HIPAA's health-care confidentiality regulations. Many workplace wellness program participants may not realize that few data privacy laws govern the information they provide, and that the vendors are frequently free to sell the data to third parties for advertising or other purposes that are unrelated to the employer wellness incentives.[27]

GENETIC INFORMATION AND WORKPLACE WELLNESS PROGRAMS

In 2008, not long after the Human Genome Project had been declared a success, Congress recognized that there were potential privacy implications in genetic information. Medical research was identifying genetic markers with diagnostic relevance: some correlated directly with certain medical conditions, while others signaled a predisposition or increased risk for various diseases or conditions. Civil rights and health advocates worried that employers could misuse this information to refuse to hire applicants with certain genetic markers, or charge increased health insurance fees to employees whose genetic makeup indicated they could be at higher risk for serious health concerns over a period of time. In response to these concerns,[28] Congress passed the Genetic Information Nondiscrimination Act (GINA), which made it illegal to discriminate against employees or job applicants on the basis of the individual's genetic information.[29]

GINA remains on the books, but individual genetic information is being used in increasingly widespread ways, many of which were likely not contemplated during the early design of the Human Genome Project and fall outside of GINA's protections. Some advocates of genetic testing are so enamored of potential benefits and uses of the technology that they overlook the associated privacy risks. As an example, in 2017 Congress considered legislation that would let employers offer lower health insurance premiums to employees who provided genetic information as part of work-sponsored employee wellness programs.[30] It's an open question whether that consent would have been voluntary in a meaningful sense, given both the financial motivation and the potential for employer pressure to participate in such programs. It's also unclear what privacy protections employers would have been obligated to place around the data, what recourse employees would have if the data was improperly shared or breached, or what obligation the employer would have had to delete the information after an individual resigned, retired, or moved on to work for another company. The legislation wasn't passed, but it's likely to resurface in some form, as employers and insurers alike are looking for ways to lower the costs of group health plans. If the preexisting conditions protections under the Affordable Care Act are undermined by future health-care legislation, it's possible that genetic information could become an increasing part of scrutiny in setting individual health-care insurability and costs.

Taken together, location tracking and health monitoring provide stark examples of the ways in which employers are able to collect and examine detailed information about the personal lives of their employees, even when those employees are off-duty, and even when the information has little correlation with the individual's ability to perform his or her job.

CHAPTER 10

DATA-DRIVEN PRIVACY DISORDER?

How Data Collection and Algorithms
Are Being Used in Education, and
What That Means for Our Kids

E dTech" is a growing field of technologies intended to assist and enhance teaching activities, student and teacher evaluation, and the overall classroom experience. Because they often promise individualized results, many of these apps and platforms also raise real and pressing privacy concerns. Data aggregation platforms are being used to enhance teacher performance, track students' moods and level of engagement in the classroom, and design more effective learning approaches, among other laudable purposes. However, these same platforms and technologies are harvesting detailed information about individual students—information that could be used to pigeonhole them into particular educational groups; to trigger, bolster, or undermine more formal educational assessments about whether they need particular kinds of education support; and in other ways impact their educational opportunities for years into the future.

GOOGLE AND AMAZON CLASS

"Our conduct is influenced not by our experience but by our expectations."

—*George Bernard Shaw*

In April 2019, *Education Week* ran an article about Google's latest classroom initiative. Google was already a major player in many aspects of education, both in informal ways like serving as the search engine of choice for students around the globe, and in formal ways like providing Google Glass as the hardware platform on which autism-therapy applications could be developed (more on that in a moment). In this particular article, *EdWeek* reported on Google's announcement that it was rolling out a new offering: cloud storage for K–12 school districts. With this new offering, school districts could store data about their students, run algorithms, and use the results to create personalized visualizations, recommendations, and predictions.[1] Washington State's Evergreen school district was one of the early adopters. The Chief Innovation Officer for the school district, Derrick Brown, explained, "We have tons of data in our school districts," including information generated by student information systems, instructional software programs, online surveys of children's social and emotional well-being, and the individualized education plans of students with special needs. Google's K–12 platform would give Evergreen a place to put all that data. And although some 80 percent of school districts were already using commercial cloud hosting through Amazon or others, most of those didn't offer the data analysis and AI capabilities that were the differentiating feature of Google's offering. The Evergreen school officials acknowledged that privacy concerns and questions about who "owns" the school district's data—students, teachers, the district, or Google—hadn't been sorted out yet. Those details didn't dampen their enthusiasm. As one Evergreen official noted, "It's a great relationship, because Google can finish my sentences. We can harness that power to learn about our learners."[2]

Google's cloud storage program is just one of its forays into the classroom. Google aggressively markets Chromebooks to schools; when a school requires all of its students to use Chromebooks, the students' coursework gets stored and indexed in Google's cloud. Students will need to set up accounts on services like Gmail and Google Drive in order to use their Chromebook, and convenience means it's likely they'll use those services for some amount of personal communications and content creation as well. As a practical matter, then, much of the students' personal data will be stored, indexed, and analyzed in the Google cloud. Whether these are e-vites for events with friends, photos taken on their phones, or a digital diary of their innermost thoughts, it's all but unavoidable that non-school-related information will make it into the students' Google-hosted accounts. And, of course, any Google searches they do while logged into their school-required suite of Google-provided accounts will be inextricably linked to the ever-expanding profile that Google creates and stores about each of them. For students who might not otherwise have created Google-hosted accounts, Google now has access to a wealth of information it would not have seen before. That information, combined with all of the students' schoolwork and school-related information, will create an even richer and more complex picture of the students' likes, dislikes, traits, and ambitions than Google could have captured if the schools hadn't required use of Chromebooks.

Google isn't alone. Amazon's voice-controlled smart speaker Echo has moved into the classroom, and its virtual assistant AI, Alexa, is being used by teachers for everything from setting timers for classroom activities to leading meditation sessions, reading to students, and, of course, bringing up music on a playlist. It isn't just Amazon that's marketing Alexa for this purpose. Third-party developers, like the startup company Bamboo (founded by a former Amazon executive), are creating interactive educational apps for students in K–12 classrooms that run on Echo/Alexa.[3]

It's hard to square these practices with the federal Children's Online Privacy Protection Act (COPPA), which requires parental consent to collect information from children under the age of thirteen.[4] In theory, Amazon allows users to request deletion of their data from Amazon's stores of information. However, in 2019, the public learned that Amazon employees were listening to Alexa-recorded voice cuts, ostensibly for quality control. Frequently, Amazon held onto the recordings for research and other purposes long after there was any plausible tie to enabling Alexa to carry out the user's command.

According to one study, Amazon's management of voice content recorded by Echo Dot, a compact version of the original Echo that Amazon markets with child customers in mind, is particularly troubling for a number of reasons. First, there's no search mechanism for parents to sort through their children's voice tracks. In order to know what the audio contains, parents have to listen to the audio or scroll through the transcriptions, making the process unnecessarily burdensome. Second, despite the requirements of laws like COPPA, Amazon has no way to verify that the parent consented to having Amazon make recordings of their child's voice. Third, Amazon keeps the recording—the child's personal information—forever, unless a parent directly and explicitly asks Amazon to delete that data. Fourth, Amazon allows a host of third-party developers to create "skills," or apps, that children can use on Alexa. Yet Amazon doesn't disclose to parents which of those third-party apps collect and keep the children's data, or what kinds of information are at stake. Fifth, when a child has a friend over, or a new child visits a classroom, Amazon doesn't give notice or get permission from parents for recording audio of the new child in the room. Sixth, Amazon's posted instructions on how to delete voice information are less than clear, as they don't explain that deleting voice recordings won't erase the associated underlying information.[5]

ELEMENTARY PRIVACY MISSING FROM ELEMENTARY SCHOOLS

All in all, researchers have concluded that "digital technology is omnipresent in elementary school classrooms."[6] Students use personal and shared devices like laptops and tablets; they use digital tools to look up information, complete assignments and projects, interact with information and with each other, and to take tests. Despite all these uses, and despite the clear presence of children online on multiple platforms, elementary school teachers are doing relatively little to educate students about online privacy or data security. The teachers who were interviewed as part of a recent study noted that it seemed as though the children were too young to need that kind of information.[7] That view seems as unrealistic as postponing sex education; if adults wait until they think it's appropriate for children to need internet privacy information, the kids will already have been misled by misinformation they've learned on the internet's virtual playground.

THE EAVESDROPPERS IN THE BABY'S ROOM

The past five years have seen dozens of news stories about privacy issues with baby monitors and what happens when hackers gain access. One hacker redirected the camera to take video of the parents' bed.[8] Another used the family's baby monitor to send threats to the family. The privacy risks range from video feeds going to Peeping Toms to hackers using the camera to gain access to broader segments of the home Wi-Fi networks, potentially including access to computers and other kinds of accounts as well as to Wi-Fi-enabled devices.[9]

Why is it so easy? According to security researchers, many baby monitors lack the basic security controls that are built into other kinds of computing devices. They can be automatically reset to factory default settings, overwriting a custom-set password; sometimes they allow security settings to be bypassed altogether.[10]

Cameras are increasingly embedded in children's toys as well. In 2017, the German government urged parents to destroy the "My Friend Cayla" doll, warning that it could be used to take surreptitious video and audio recordings of children.[11] The doll can be controlled by an app and uses speech recognition software to allow children to access the internet. But these features—while carrying a certain "gee whiz" panache in the United States—prompted German regulators to declare the doll an "illegal espionage apparatus" under German laws that prohibit surveillance devices from being disguised as other objects.[12]

The German regulators' ban brought to light the tension between privacy protection laws and the individual consumer preferences. On the one hand, Cayla was almost completely unprotected: it could be activated without a password, and its Bluetooth signal could be accessed from fifty feet away, allowing a hacker to eavesdrop on the child's conversations with the doll and even talk to the child through the doll's embedded speakers. However, it also had wide commercial appeal: Cayla had been named a "top ten" toy of 2014 by a German toy trade organization.[13] When Cayla ran afoul of German law, it was illegal to make, sell, or *possess* the doll. In theory, regulators could prosecute families simply for having a Cayla doll in their homes—an object that, in any given family, might be the child's favorite toy.

Examples abound of poorly protected toys that connect to the internet. These include Cloud Pets, which were pulled from stores after security researchers discovered that children's names, ages, and voice recordings were all easily accessible on the internet, because the manufacturer had failed to implement any security controls on the cloud system that it used to store records—including audio files—from and about the children using them.[14] In a different kind of privacy incident, the 2015 cyberattack on the network of toy company VTech allowed hackers to gain access to the home addresses and photographs of more than six million children.[15]

In recognition of these mounting concerns, the FBI issued a statement in 2017 warning about the privacy risks of internet-connected toys:

> The FBI encourages consumers to consider cyber security prior to introducing smart, interactive, internet-connected toys into their homes or trusted environments . . . These toys typically contain sensors, microphones, cameras, data storage components, and other multimedia capabilities—including speech recognition and GPS option. These features could put the privacy and safety of children at risk due to the large amount of personal information that may be unwittingly disclosed.[16]

According to the FBI statement, these toys create risks of "child identity fraud" and other, more chilling, consequences. In some cases, toys that haven't been activated can record conversations within earshot, gathering information about children's names, schools, likes and dislikes, and activities, along with past and real-time physical locations, internet use history, and internet addresses. As a result, "the potential misuse of sensitive data such as GPS location information, visual identifiers from pictures or videos, and known interests to garner trust from a child could present exploitation risks."[17]

As with other areas of life, privacy intrusions in schools put other values and interests at risk as well. In the case of technology-based school instruction, one potential consequence is a move away from the personal growth, interpersonal feedback, and face-to-face interactions that are so necessary to forming a sense of who we are. Software companies have promoted online pre-kindergarten programs as an alternative for students in communities where pre-K isn't funded by local governments. It's hard to estimate just how great the data privacy risks are for a program in which all student interaction is online.[18] Nonetheless, the proposal has garnered attention through TED talks and the TED-backed The Audacious Project, with supporters praising it as an innovative way to close the gap between haves and have-nots. Critics worry that it could divert money from in-person classrooms. And education advocates don't yet know how to gauge the impact on values of personhood. A lack of autonomy, emotional release, and the ability to have a one-on-one conversation with a trusted teacher could result if the availability of tech-only teaching for young children becomes a model that begins to replace in-person education.

VIOLENCE PREVENTION THROUGH SCHOOL SURVEILLANCE

Despite uncertainty about how to safely and effectively use data-driven technology in the classroom, a number of schools are hoping technology will help them increase safety by identifying potential school shooters and creating opportunities to intervene earlier with students who are feeling disgruntled or disengaged. One Canada-based company is offering facial recognition technology to schools to identify individuals who aren't authorized to be on campus, and to identify images of guns if someone is walking the halls of a school openly carrying one. It's unclear precisely how this technology would help provide early warning to schools—particularly because so many school shootings are carried out by current students who would presumably be recognized by the software as being authorized to be on the premises. But school districts like Lockport, New York, are investing heavily in the technology in hopes it will make students safer.[19] Although the school district only has about 4,500 students, the district has spent some $3 million of a $4 million Smart Schools grant on expenses like adding or upgrading 417 cameras to capture high-resolution photographs. Some parents have objected to the plan, for both privacy and efficacy concerns; this program is one where the privacy intrusion is likely to be high, the usefulness may be low, and the economic impacts all too real and significant.[20]

Although Lockport is the first to try this particular product, there are a host of AI tool suites that promise to identify students who pose risks to school safety. The business model has become so widespread that a major consortium of education and advocacy groups have issued a set of guiding principles intended to prevent the most vulnerable students from becoming targets of adverse action based on a combination of profiling and paranoia.[21] While many school safety proposals call for increased surveillance as a tool for decreasing school violence and self-harm, privacy advocates argue that the surveillance can disproportionately affect students of color and students with mental health issues.[22] These concerns are well founded: in some cases, schools are explicitly targeting students with mental health concerns. Florida, for example, passed a 2018 law requiring schools to ask about previous mental health referrals as part of school registration, and proposed that students who have documented behavioral issues on an individual education plan be referred to and evaluated by the school's threat assessment team.[23] In the wake of the Newtown, Connecticut, school shooting, research has shown that schools whose student body was majority non-white were more likely to see increased measures like x-ray machines, armed guards, random searches, locked gates, and security cameras—all measures that have questionable value in preventing school shootings and that contribute to an overall atmosphere in which students are treated more like inmates than like students, and in which maintaining a climate devoted to learning becomes more difficult to achieve.[24]

DATA-DRIVEN MEASURES OF STUDENT ENGAGEMENT

US schools rely heavily on student questionnaires to measure their "social emotional engagement." But digital technology may allow schools to turn to other, more intrusive means, such as using facial recognition technology to assess whether or not students are paying attention. In experiments around the world, artificial intelligence is being paired with video surveillance to measure student "engagement" in the classroom. The resulting data is incorporated into achievement scores for both students and teachers. This new path was paved by a school in eastern China that installed three FRT-equipped cameras in 2018. The cameras could be used for routine purposes like checking attendance in the classroom. But they were also used to monitor the behavior of students and of the teacher, assessing the effectiveness of the teacher's approach and gauging the students' "engagement" in the classroom.[25] The cameras, which formed part of the school's "Smart Classroom Behavior Management System," scanned students' facial expressions and apparent level of attentiveness every thirty seconds. One student quoted in news reports gave the initiative positive reviews, saying, "Beforehand in some classes I didn't like much, sometimes I would be lazy and do things like take naps on the desk or flick through other textbooks. Since the school has introduced these cameras, it is like there are a pair of mystery eyes constantly watching me, and I don't dare let my mind wander."[26] A skeptical reader could inquire how free a student might feel to express a critical opinion of the new cameras in a society that's leading the world in government surveillance of its populace.

> *Artificial intelligence is being paired with video surveillance to measure student "engagement" in the classroom. The resulting data is incorporated into achievement scores for both students and teachers.*

Regardless of whether that particular student was genuinely a fan of this approach, the use of FRT in attempts to measure student engagement isn't limited to one classroom, one school, or one country. The University of Montreal is partnering with private company Classcraft to develop new technology-driven measures for student engagement.[27] Separately, a business school in Paris has been using a software called Nestor since 2017 to try to measure student engagement.[28] The technology operates through students' webcams and measures movement across twenty key landmarks on each student's face—eyes, brows, lips, jaw—to draw conclusions about whether the student is paying attention.[29] The company won't retain the data, but the school will integrate the student's FR scans with other online activity, like calendars and social networking platforms, to get a fuller picture of the student's behavior and even make recommendations about when the student should study. (If, for example, Nestor notices that a student watches YouTube videos on Sunday evenings, then Nestor might send the student a prompt nagging them to study instead.)[30] According to some accounts, students had no opportunity to opt out of having this technology track their behavior during their online class sessions. And Nestor deliberately incorporates a "gotcha" model: its algorithms identify moments when students appear unengaged, and then asks them questions from that portion of the course material. According to the company's founder, "It's impossible to pass the exam if you're not 200 percent focused." Despite the Nestor founder's confidence, it's unclear whether the technology works in the way that it's advertised. It achieves the goal of measuring changes in facial expression. But according to researchers who assess learning outcomes, there's very little correlation between whether a student *looks* engaged and whether they are actually absorbing information.[31]

The use of AI in education is expected to grow 47.8 percent between 2018 and 2022.[32] In K–12 alone, that growth is expected to be fivefold in seven years, from $200 million in 2017 to $1.2 billion in 2024.[33] Meanwhile, student data—the fuel for all of that artificial intelligence—is rarely deleted or aged off. Privacy advocates are turning their attention to the problem of student data that's gathered once, lives forever, and multiplies seemingly infinitely.[34] Some data, like report cards, have utility over a long time frame. But many other kinds of student data are highly perishable, losing any usefulness shortly after collection.

With the growth of classroom technology in mind, a Maryland father proposed establishing June 30 as "National Data Deletion Day." The aim is to draw attention to, and correct overlong retention of, data that has no educational value when kept for a long time but that can infringe on student privacy. Under this proposal, every year schools would have to delete information like students' internet browsing history and emails, their online class interactions and assignments saved on cloud-based platforms, location information from digital devices, and biometric information like fingerprints used for meal purchase accounts. The father who proposed Data Deletion Day explained his motivation: previous generations had the freedom to pursue all manner of interests, some of them fleeting, without that information being sold to advertisers or harvested for algorithmic review by the school to form part of a perpetual student file.[35]

As digital data about students expands, it remains to be seen whether privacy-protective tools and privacy-protective rules will grow at the same rate. In the meantime, it's worth asking the question: If we've lost the ability to gaze out a classroom window and daydream, to research new topics or to stare thoughtfully into space—in other words, if we're losing the privacy of our own thoughts—how much privacy can we say that we have left?

CHAPTER 11

WHEN YOUR DATA IS YOU

Facial Recognition, Biometric
Technology, and Public Health

C ompanies like Walmart are using facial recognition data to clock employees' arrival and departure times from work. The US Department of Defense has been looking at gait recognition as a means of identifying individuals since at least 2002. A recent Georgetown University report estimated that approximately 117 million Americans are currently identified in facial recognition databases. Some of this unseen collection of biological information could offer promising individual or public health benefits. But some of it can be downright dangerous to autonomy, liberty, and freedom from discrimination. Data that's intrinsically tied to our biology has the potential to be uniquely damaging; we can't, after all, change these fundamental features of who we are. This chapter starts with some promising innovations, and then examines some troubling examples that raise the question: Are there any limits to how employers, schools, retailers, and others can collect, access, and use that data? Should there be?

SMART TOILETS AND AI DIAGNOSTICS

Biometrics are largely thought of as physical characteristics that can be used to identify an individual. Given the range of personal data available, however, it's also useful to look at the unexpected ways in which private technology can capture data relating to our body's functions, and combine that with other information to connect our individual identity with information about possible health conditions—in some cases, conditions that neither we, nor our doctors, know that we have. The possibilities are nearly endless, but two examples—smart toilets and AI-driven diagnosis—help illuminate what's currently happening, and where the technology might go in the future.

A handful of companies and biomedical researchers have work well underway to use toilets equipped with digital sensor technology to analyze human urine and waste for a variety of diseases, conditions, and indicators of personal habits. By 2021, smart toilets could be both readily available and commercially viable. The TrueLoo, for example, is a smart toilet whose optical sensors have the potential to detect dehydration, various viruses, and urinary tract infections from the user's urine. TrueLoo's maker, ToiLabs, plans to market the smart toilet to senior living facilities in the San Francisco Bay Area in 2020 and make it available to consumers not long after that. Meanwhile, a group of researchers at the University of Madison, Wisconsin, have run tests using urine samples to detect a number of lifestyle attributes, including the data subject's nutrition, ability to metabolize over-the-counter drugs, sleep patterns, and how often the subject exercised and how much caffeine they consumed.[1] Whether smart toilets become common residential features, or their use remains limited to settings like hospitals and nursing homes where the diagnostic benefits can be particularly valuable, the upcoming decade appears likely to herald a breakthrough in the creation of smart toilet technology that is sufficiently low-cost enough to come into widespread use.

If the diagnostic benefits of smart toilets seem a bit earthy or gritty, the latest salvo in Parkinson's diagnosis provides a second example at the other end of the data-gathering spectrum, based almost entirely on a user's interaction with computer devices. In 2018, Eric Horvitz, a Stanford MD and PhD and head of Microsoft's research labs, announced that, together with his research team, he had devised a set of algorithms that could provide early detection and diagnosis for people who would develop Parkinson's disease. The keys to the algorithm are the digital tracks created by a cursor: the ways that a person with nearly imperceptible tremors moves a computer mouse across a desktop. The diagnosis isn't based on physical movements alone; it combines mouse control information with data about the kinds of web searches the computer has carried out. Horvitz has developed other algorithms as well: using an individual's search history data to identify users with pancreatic, lung, and breast cancer, and examining millions of anonymized search history logs to assess the impact of sleep on performance by evaluating the relationship between a user's apparent sleep habits and the speed of their keystrokes and mouse clicks.

The heart of all of these diagnostic tools is the artificial intelligence algorithms that review information like times and dates of searches, keystrokes, clicks, and mouse activity, harvested from the anonymized data of hundreds of millions of web searches. Horvitz envisions a day when a set of algorithms running somewhere in a computer's background could detect the anomalous mouse movements, compare them with relevant internet search data, and notify the user that they have the disease, encouraging them to see a doctor to discuss treatment options. The idea is that, in the future, users could opt-in to these services and agree to have their searches and physical interactions monitored, and then be given prompts alerting them to potential medical conditions.[2] Even if the data is anonymized, however, it raises thorny questions for medical ethicists, such as how to obtain informed consent, how to protect the information from misuse, and what standards would govern how the data can be used.

From smart toilets to mouse tracking, these technologies offer plenty of promise and also raise a host of questions. What if a workplace installs smart toilets as part of a wellness program but doesn't tell its employees—either that it's collecting the data or that it might try to associate that data with individuals as part of assessing their overall health risk? Alternatively, suppose that an employer installs mouse tracking as part of an overall set of insider threat and cybersecurity monitoring tools. Workers know that their employer has the right to read the emails they send from their work address, and to monitor the websites they visit from their work computer. But what about employers making use of that data to assess the likelihood that a member of their staff could develop Parkinson's in the future? Does the employer have an obligation to tell the worker that its computer monitoring tools could reveal certain medical conditions, or that a specific individual may be at risk?

WHEN YOUR IDENTITY IS USED TO MARKET—AND TO DISCRIMINATE

In 2018, a controversy erupted in Canada over press reports that a mall was using facial recognition technology to assess the age and gender of shoppers and to track their movement through the mall. Even though the mall operators said they were not capturing individual images, the fact that the technology was being used without notice or consent from shoppers prompted complaints from the Canadian Civil Liberties Association and an investigation by the Canadian Privacy Commissioner's office.[3] Shopping malls are by no means the only private-sector entities using facial recognition for surveillance; and monitoring traffic patterns is by no means the only use. Celebrities like Taylor Swift, who have been targeted by stalkers, are using facial recognition systems at their concerts and events to look for any one of the hundreds of individuals who have stalked the artist in the past.[4] It's unclear who owns the rights to the images collected by these systems, or how long Swift's security team keeps the footage, but—with the exception of a handful of state laws—the use of FRT by private-sector entities is largely unregulated in the United States. Many consumers welcome the convenience promised by FRT: Ticketmaster recently invested in Blink Identity, a startup company claiming that its FRT can identify individuals while they're walking at a normal pace, something that could presumably be used—like a human EZPass—to move concertgoers through turnstiles more quickly than having human ticket-takers scan barcodes or rip off the stub on a paper ticket.[5]

Companies selling FRT systems often base their sales pitch on its uses for worker identification and business security. Compared with other identification systems, they claim, it's the "least intrusive method that provides no delays and leaves the subjects entirely unaware of the process."[6] It's being marketed as a way to spot shoplifters, "scam artists or potential terrorists," and "VIP customers who need special attention."[7] Of course, a facial recognition system can only identify specific individuals if the system is trained with images of known desirable persons (such as VIP customers) and undesirable persons (previously apprehended shoplifters, former disgruntled employees, and the like). That very significant limitation on effectiveness is rarely noted in marketing brochures.

The private sector is looking at using facial recognition in a number of other ways. As if straight out of a dystopian science fiction film, Microsoft has patented a billboard that can identify individuals as they walk by and serve up individually tailored ads.[8] In order for this system to work, Microsoft needs FRT scanning; rapid image storage; a vast catalogue of previously collected and identified images, annotated with the person's likes, dislikes, shopping habits, and income; and near-real-time analytics to compare the new image against its database. Retailers in China have been using facial recognition that lets customers "smile to pay" since 2017.[9] And a 2017 editorial in the *Financial Times* pointed to the risks of differential pricing: that is, the risk that retailers would use FRT to identify specific shoppers, or to read their moods or appraise likely socioeconomic class, and offer different prices to each shopper depending on what the facial recognition algorithm predicted the person would likely be willing to pay.[10]

FACIAL RECOGNITION AND THE RIGHT TO BE LEFT ALONE

In 2016, news broke about an app called FindFace that allows smartphone users to take a snapshot of a stranger, upload the photo, and find out the person's name and other details. The app, developed in Russia, compares the new image with nearly a billion photos on widely used social media sites in the region. In its first two months, FindFace attracted half a million users who performed nearly three million searches.[11] Although many users likely downloaded FindFace for fun, Russian law requires all personal data collected in that country to be stored on servers that can be accessed by the government. It's not hard to imagine that Russia's current repressive regime might use FindFace data to identify protesters and dissidents, squashing freedom of association, speech, and debate.

All of these technologies, however, are equally susceptible to misuse by private entities, reflected in the fact that FindFace app makers proposed some uses that were downright creepy. One of the app's developers predicted that it could revolutionize dating: "If you see someone you like, you can photograph them, find their identity, and then send them a friend request. It also looks for similar people. So you could just upload a photo of a movie star you like, or your ex, and then find 10 girls who look similar to her and send them messages." Creepy commercial use was also in-scope: per the developer, stores could take photos of customers looking at particular merchandise, identify them through the app, and then send them marketing emails as follow-up to their in-store browsing.[12] A whole host of vigilante surveillance could result. Anti-abortion activists could use the app to identify patients or employees entering clinics and send them harassing emails. Anti-gay activists could use it to identify people entering gay bars and send them harassing messages. Religious bigots could use the app to identify people entering synagogues, mosques, or churches, and send them threats. The list of potential abuses is a long one—and, unlike government use of facial recognition, when these kinds of activities are carried out by private individuals, there is no oversight framework in place to prevent the app from being misused.

So just how intrusive is it to have employers, fellow commuters, store owners, and passersby watch and record our every move? We don't know yet because we don't know what limits this technology might eventually reach, or what regulations might be passed to govern it. In the United States, some judges and lawmakers have started to recognize these concerns. The Illinois state legislature passed the first biometrics protection law in the country over a decade ago, but the law received renewed attention when the Illinois Supreme Court ruled that individuals could sue companies who collected and used their biometric data without permission. In a string of cases since then, other courts have also upheld this right, including a landmark decision from the federal Ninth Circuit Court of Appeals, which noted that Facebook's facial-recognition technology can obtain information that is "detailed, ency-clopedic and effortlessly compiled." Further, it was intrusive: "Once a face template of an individual is created, Facebook can use it to identify that individual in any of the other hundreds of millions of photos uploaded to Facebook each day, as well as determine when the individual was present at a specific location. Facebook can also identify the individual's Facebook friends or acquaintances who are present in the photo."[13] According to the court, this was precisely the type of privacy risk that Illinois lawmakers had sought to protect against; consequently, the lawsuit against Facebook could go forward. In the wake of that ruling, Facebook agreed to pay $550 million to the class of Illinois residents in compensation for its use of facial recognition without first gaining their consent.[14]

In addition to the direct impacts recognized by the Ninth Circuit, it's also worth considering the way that widespread use of FRT could undermine our overall sense of freedom to move unobserved through society, and to live unencumbered by the expectations that come with being watched. In China, a customer-service company allows staff to wear masks in the office on No-Face Day as a reprieve from the daily pressure to be unfailing in car-rying out each customer interaction with a smile. The name of No-Face Day is based on an anime character with a blank face and an inscrutable expres-sion.[15] With facial recognition cameras hidden in malls and connected to street surveillance cameras, with our photos being catalogued by governments and searched and identified by anyone who's ever had an account on social media, we might all find wearing No-Face masks to be a relief.

WHEN BIOMETRIC DATA IMPACTS YOUR STANDING IN SOCIETY

This tension between the benefits of personalized DNA profiles and the privacy challenges associated with it arises in unexpected ways, and has the potential to extend across nearly every sphere of life.

According to the *MIT Technology Review*, by early 2019, more than 26 million people had provided DNA samples for at-home ancestry tests.[16] MIT estimated that, based on current trends, that number would likely increase to more than 100 million people within two years. The companies offering at-home ancestry kits are privately owned and they're not considered health-care providers—meaning their obligations for transparency, disclosure, and review of their data handling practices are, under current laws, inherently lesser than might be the case for companies that are subject to regulatory scrutiny.

The privacy implications of DNA testing are both quantitative and qualitative. To start with the "soft" concerns: the internet is replete with anecdotes about individuals who used an ancestry/DNA kit and subsequently discovered that their family history was quite different from what they grew up believing. Ancestry.com's privacy statement acknowledges this risk, warning users that "You may discover unexpected facts about yourself or your family when using our services. Once discoveries are made, we can't undo them."[17] That's led to some odd exchanges in chatrooms frequented by neo-Nazis and other white supremacists, saying that even if a home DNA test showed their heritage to include a significant percentage of, say, sub-Saharan African ancestry, it doesn't matter. Either they reject the tests entirely or start to make bargains about what percentage of non–northern European heritage might be acceptable. It seldom causes them to rethink their white supremacist ideology—at least not openly, on websites that are havens for those views. In effect, many of them say, "I don't care about any of those tests. I *feel* white."[18]

What we know from the experience of just the past few years of voluntary genetic testing by private companies is that once an individual has given up their spit swab, they have little to no control over how the company they provided it to will use it: whether and how it will be disclosed to others (potential relatives as well as law enforcement or other government officials) or be combined with other kinds of information (for medical research studies, genealogical research, or other purposes).

This brings us to a quirky new use of DNA testing: policing the actions of pet owners who don't clean up after their dogs. PooPrints is a scientific company based in Tennessee, which, for about $65, will carry out lab testing to look for strands of DNA in samples of dog poop, and compare any DNA they find against the DNA World Pet Registry, their proprietary database of known dog DNA. Their customer base consists of gated communities, apartment buildings, and other enclaves where residents share a community-owned space for pets to relieve themselves—and pet owners don't always bother to pick up after their dogs. PooPrints opened up shop in 2008, and a decade later, their customers included over three thousand communities in the United States, Canada, and the United Kingdom.[19]

By coincidence, I learned about PooPrints right after watching a BBC News video about the uptick in communities installing cameras to record the license plates of vehicles passing through. Perhaps the visions that video triggered of overzealous community members using technology for private policing weren't so speculative or outlandish after all. It turns out that a private community in Maryland, not far from where I live, has imposed new rules: all pet owners must turn over cheek swabs of their dogs' spittle so that the animals' DNA can be tested. Once the dogs' DNA has been entered into PooPrints' database, any errant poo can be matched against the offending pooch forever into the future. Armed with the dogs' DNA, the next time a property manager or member of a homeowner's association finds poo in a place it shouldn't be, they can scrape a sample into a vial and ship it to the company's lab in Tennessee. There, the poo is swirled around in a centrifuge that isolates the strands of DNA contained in the waste. The isolated DNA is then compared against the known DNA samples of other pets registered with the service. PooPrints notifies the community when they find a match, with what the company estimates as less than a one-in-one-quintillion chance of error. With the culprit identified, the community decides what action to take, presumably including some combination of warnings, fines, or even eviction of the owner if the violations are regular and serious enough and if compliance with these conditions is written into the property deed or lease.[20]

There's nothing unreasonable about communities wanting residents to clean up after their pets. And it's hard to argue that a dog has a legally protected privacy interest in their own DNA. And yet, whether the scenarios are as unsympathetic as the disappointment of white supremacists who wish their DNA were more white, or as light-hearted as the testing of dog poo DNA, the increasing availability of uniquely identifying DNA is just one of the many pieces of the digital data privacy puzzle that puts personal relationships and social standing at risk.

To take a more extreme example, the government of China encourages residents to turn over information about their neighbors' missteps, all part of the process of expanding the massive national database of individuals' social credit scores.[21] Perhaps this desire to judge is more than a quirk of human nature, but an innate feature: the desire to set standards for acceptable behavior and hold others accountable to living up to them. Although many of us say, "Live and let live," all too often, we don't really mean it. The idea that privacy entails the "right to be left alone" resonates with us in the abstract. But oftentimes, what we really mean is, "I'd like to be left alone and I'm willing to leave others alone, too, so long as they are conforming with the particular societal expectations that are most important to me." Whether that's catching drivers going 32 mph in a 25-mph zone, or upping the ante on neighbors who fail to clean up after their pets, public shaming and approbation are important, effective, and time-honored tools in maintaining a sense of community order, whether for ill or good. What's changed with newly available and ever-cheaper technologies is that the opportunities for private citizens to know what their neighbors are doing are becoming ever more widespread and available. As data-driven technologies undermine our ability to act anonymously and to be left alone, the old adage holds that good fences make good neighbors. It remains to be seen whether advancing technology does the same.

BIOMETRIC IDENTIFICATION AND PUBLIC HEALTH

In early 2020, public health officials around the world were grappling with the spread of a new and deadly form of coronavirus that had originated in Wuhan, China. Authorities in Wuhan were widely criticized by the global health community for what appeared to have been a slow response to the virus, downplaying early concerns, and providing the public with false reassurances.

Those criticisms came to a head when Dr. Li Wenliang, an ophthalmologist in Wuhan, died on February 7, 2020. His death unleashed a torrent of anger and grief throughout a country where public dissent and criticism of the government are normally both heavily censored and rare.

Dr. Li had been among the first to treat the virus and warn about its dangers. He was also one of the most vocal critics of government disinformation that downplayed its risks.[22] On December 30, 2019, Dr. Li had shared a post, via a social messaging app, with a group of his medical school classmates suggesting that this new virus might become a threat. The response from government censors was swift and severe: Dr. Li was forced to publicly recant his statement. The document he was told to sign asked, "Can you stop your illegal behavior?" and "Do you understand you'll be punished if you don't stop such behavior?"

In the end, his warnings proved to be correct. As a result of the outbreak, China's public health authorities instituted an unprecedented quarantine around the entire region surrounding the city of Wuhan, imposing a strict travel ban on some fifty million people from the affected areas.[23] As part of its overall efforts to curtail travel by potentially infected individuals, China deployed thermal imaging sensors at rail stations and airports around the country, using infrared technology to ferret out travelers who were running a fever.[24]

In the midst of the quarantine, China's state media outlet, the *Global Times*, posted a video that quickly went viral. "Can't stop farting?" the headline asked. "Well, you better try."[25] It purported to show the heated emissions of farts being expelled by people and dogs in busy transit stations, all of whom were being monitored by the thermal imaging being used to check for traveler fevers. The puffs of flatulent emissions depicted on the video were a hoax.[26] However, it was also a powerful example of disinformation: within days, major news outlets were pointing to sources as varied as digital forensics of the video and past episodes of *Mythbusters* to debunk the video.[27] If thermal imaging were able to detect the heated air of a fart, it would also show the heated emissions of an exhale each time a person breathed. What, then, was the point of the doctored video? It's hard to say. Perhaps the goal was twofold—to offer some seeming levity in the midst of a crisis, while simultaneously reminding China's residents of a reality that was all too familiar to them: the ubiquity of state surveillance.

Before posting its viral fart videos, the *Global Times* also posted video clips of drones berating people who were out in public without the face masks that Chinese residents had been urged to use to combat the spread of the disease.[28] The videos showed a drone flying over a city intersection with a loudspeaker that harangued a group of women crossing the street without masks.[29] Another clip showed an aerial view of a man in a rural setting in Inner Mongolia, climbing onto a tractor without a mask; the loudspeaker berated the man and told him to return home: "Uncle, we are in unusual times . . . Now get on your cart and go home immediately."[30]

What ties these stories of quarantines and drone surveillance to biometrics? The widespread net of surveillance sensors across China collects and compiles data that provides detailed information on precisely who might be trying to dodge public health restrictions. When a Hangzhou man returned from a business trip to an area that had experienced a coronavirus outbreak, local authorities asked him to remain at home, on an informal quarantine, for two weeks. Twelve days later, bored with inactivity, the man left his house. He was promptly contacted by his boss and by police: facial recognition cameras had spotted him by Hangzhou Lake.[31]

The Chinese government also worked with the popular financial services company Alipay (sister company to retail behemoth Alibaba) to roll out a mobile app that would allow residents to upload their location and identification number. The app would then calculate the user's level of contagion risk, based on whom they had recently been in contact with and where they had been. Very little information is currently available about how the algorithm reaches its conclusions—what factors it weighs, what weights they are assigned, what other information it might draw on. From that calculation, the app then assigns the user a color code—red, yellow, or green—indicating whether they should be in quarantine. According to researchers, that color code is also reported to police, who can then enforce the algorithmically generated quarantine recommendation if the person fails to follow the required actions associated with their assigned code.[32]

Other companies have advertised new technologies that combine biometric identification techniques with temperature detection, which could be deployed at building entrances to determine the specific identity of individuals who enter a building while running a fever. Chinese public health authorities view these capabilities as an improvement over the SARS outbreak in 2003, when it was harder to track the movement of individuals who might have been infected or exposed to the disease. Officials in China predict that they can now better identify sick individuals—and anyone who has been in contact with them—and provide the information to "epidemic prevention departments."[33] And consequences for individuals who are caught defying these government edicts can be severe. By February 2020, infected individuals were being forcibly moved to quarantine camps, and health officials were interviewing everyone they had come in close contact with. The Chinese vice premier responsible for leading the response to the outbreak was quoted as saying, "Set up a twenty-four hour system. During these wartime conditions, there must be no deserters, or they will be nailed to the pillar of historical shame."[34]

Data-driven technologies provide governments with more tools than ever to track, influence, and even control the behavior of the people who live within their borders. The risk of abuse is often greatest in times of crisis, whether that's a national security crisis, a public health crisis, or some other threat. To the extent that decisions having real-world impact on individuals are made by black-box algorithms, the lessons going forward are likely to be mixed: the power of big data to make recommendations, the dangers of allowing algorithms to make decisions automatically, the distrust that fosters when there's little transparency about how the algorithms work and no recourse for their decisions, and the likelihood that after-the-fact review will reveal mixed results across the spectrum of accuracy.

The Chinese government's overall program of internal security surveillance will be addressed in more detail in chapter 20, describing the ways that authoritarian regimes use personal data to shape political messaging, cement their power, and exert control. For the purposes of this chapter, however, the Chinese response to the coronavirus has provided an object lesson in the ways in which this data is combined with biometric information, ostensibly in the service of public health but, if carried out with little regard for liberty, at considerable expense to the individual.

DIGITAL DATA AND COVID-19

Questions about digital information, privacy, and public health have become increasingly urgent as the novel coronavirus pandemic spread across the globe from late 2019 through 2020. As this book was going to print, millions of people around the world had contracted COVID-19, hundreds of thousands had died, and by some estimates roughly half of the world's population—some four billion people—were under some sort of stay-at-home order.[35] Governments worldwide were examining ways to use digital surveillance for public health: South Korea was tracking citizens' movements through mobile phone data, credit card purchases, and their cars' GPS; Israel's Shin Bet security agency was revamping terrorist-tracking tools to do contact tracing and measure social distance; China repositioned surveillance cameras to monitor the doorways of infected residents.[36] In the United States, local governments were considering programs that would use drone surveillance to monitor social distancing in public places, warn people to go home, and even measure their body temperatures using infrared sensors.[37] Google and Apple announced an app that would use Bluetooth signals to notify cell phone users if they came in physical proximity with a phone owned by someone who had been infected with COVID-19.[38] Meanwhile, the Food and Drug Administration issued guidance lauding the value of digital technologies for public health—and disclaiming any responsibility for regulating them, as they are not "medical devices."[39] At the same time, right-wing groups in the US were staging armed demonstrations protesting government stay-at-home orders,[40] creating a potentially volatile mix of triggers for privacy intrusion, as public health and law enforcement officials seek ways to use technology to monitor protests, surveil domestic terrorist groups, carry out contact tracing, and enforce social distancing, all in hopes of achieving the complex set of public health metrics and public safety conditions that might make it possible for businesses to reopen and for people to safely return to work.

The good news: digital data greatly expands the ways that individuals, organizations, and governments can gather the information that's needed to understand the novel coronavirus, to slow its spread, and to measure the effectiveness of public health interventions designed to combat it. The bad news: countless other circumstances have shown that it's easy, and tempting, to cut corners on privacy during a crisis. The challenge is to make the best possible use of the tools that are available—and to make sure that that policymakers, businesses, and individuals all understand that the "best possible" uses are ones that incorporate thoughtful decisions about privacy alongside utilitarian economic or public health assessments of what new algorithms, apps, and data collection programs can do.

CHAPTER 12
UNDERPAID DATA LABOR

AI Training, Digital Piecework,
and the Survey Economy

Artificial intelligence and complex algorithms require lots of data, and lots of people to train the data. It turns out that all of us have been unwitting participants in a multifaceted, loosely designed program of unregulated research. This initiative has an undisclosed threefold effect of extracting our personal information, training AI algorithms, and providing a source of nearly cost-free labor that supports the continuation of the first two effects. This trio of interrelated activities impacts our privacy across multiple dimensions: gathering information we might not otherwise disclose and sharing it with others who will benefit from the immediate monetization as well as future opportunities to use that data for microtargeting; undermining our sense of autonomy; and calling into question what personhood means in a society where algorithms are becoming ever more effective at mimicking human interactions and behavior.

This triangulation started with the Turing test, evolved into CAPTCHA boxes, and proceeds through online surveys.

FIRST, THE TURING TEST

The test is named after Alan Turing, a brilliant crypto-mathematician who worked for the British government during World War II at Bletchley Park. He devised the computer that enabled Allied forces to crack the German codes and through that staggering achievement helped turn the tide of the war. Despite his extraordinary contributions to science and national security—literally, to the British victory in the war—Turing faced criminal prosecution after the war for engaging in homosexual acts. He was stripped of his security clearance, barred from working for British intelligence, and sentenced to chemical castration. He died at the age of forty-one from cyanide poisoning, in what was ruled an act of suicide.[1]

It was Turing who first proposed, in 1950, a definition for artificial intelligence, inspired in part from a test that later became known as the "Turing test." The Turing test was based on a parlor game similar to twenty questions, where the goal was for a person sitting in a room by themselves to figure out, gleaned from written answers to questions, which of the two people in an adjacent room was male or female. In the flesh-and-blood version of the game, the man tried to masquerade as the woman, the woman answered as herself, and the questioner in the other room had to figure out which was which. In Turing's version, the deception was switched: a computer could be said to "think" if, in that question-and-answer exchange, the real man and woman couldn't tell that the entity asking them questions was in fact a computer rather than a person. In its simplest terms, a computer passes the Turing test if it can fool the people it interacts with into thinking that it's human, and not a piece of software running on electronic circuits.[2]

In creating the test, Turing's goal was simple: "I propose to consider the question, 'Can machines think?'" In answering that question, Turing made startling, and visionary, predictions:

> I believe that in about fifty years' time it will be possible to programme computers . . . so well that an average interrogator will not

have more than 70 per cent chance of making the right identification after five minutes of questioning.

Turing noted that the "only really satisfactory" way to answer his question was to wait fifty years and see how computers had developed. Nonetheless, he expected that by the end of the twentieth century, the sophistication of computer programs would have increased so much that it would be commonplace for people to talk about machines being able to "think."[3]

As Turing predicted, by the end of the twentieth century, there was an entire field of computer science dedicated to constructing tests to measure a computer's "artificial intelligence" by the standard Turing had described: whether a computer could masquerade so convincingly that it fooled people into thinking it was human.

THE CAPTCHA TEST

If you've spent any time on the internet, you've undoubtedly navigated to a site demanding that you click on a box saying "I am not a robot" before it would allow you to proceed. Sometimes the program presents you with a grid of blurry pictures and asks you to pick out all of the ones that include storefronts or cars; sometimes you have to correctly enter a sequence of numbers and letters perceptible in a distorted image.

All of these are part of a system of digital testing designed to prove whether the entity requesting access to the website is a human or an automated software script. The websites that ask for this are typically doing it because they don't want to be overrun by traffic from bots that might carry malicious code. The websites use CAPTCHA, which stands for Completely Automated Public Turing Test to Tell Computers and Humans Apart, to present you with a question or task that people are good at performing but computers are not. In other words, they ask you, the user, to pass the Turing test.

These CAPTCHA tests proved so effective that Google bought the software in 2009.[4] At the time, Google was digitizing its library of online books and needed humans to validate the text it was digitizing. As computers were attempting to identify the words in text of different fonts and varying degrees of legibility, Google needed humans to verify that the computer had gotten the text right. Why? Because this validation was an essential part of the process of training AI algorithms to do their job. It was a brilliant move: buying CAPTCHA allowed Google to profit from multiple stakeholders in a single transaction. Google got free human labor to validate the digital text and train its algorithms. Websites that were being indexed by Google wanted a way to ensure that the traffic directed to their sites—often by Google—was human traffic, not bots. Google was able to further profit from providing a security check that became another part of its suite of services. And the humans asked to perform these tasks didn't object because each specific request took such a small amount of effort and time.

Just as Turing had predicted, the Google AI became steadily more accurate at recognizing text, and by 2013, the renamed security check (now called reCAPTCHA) needed new challenges. Google's AI began evaluating at mouse usage, site navigation, and other features of physical and virtual interaction with a site to predict whether a user was likely to be a human or a bot. When the AI concludes that the user is probably a human, then the "I am not a robot" checkbox appears, allowing the user to check the box, press "Enter," and proceed directly to the website they want to access. In cases where the user's behavior seems anomalous, reCAPTCHA began presenting something new: a challenge box consisting of photos that the user had to tag. At first the photo challenges were simple ones: pick out the photo of a cat. Clicking on the cat served the security purpose, because most AI wasn't good enough at the time to identify specific objects in photos. It also served Google's purpose, because each click trained Google's AI to get better at recognizing photos of cats. As the AI got better at recognizing photos, the test became harder, leading to the grainy and indistinct pictures that are so commonly used in the reCAPTCHA grids today.

By 2019, AI researchers were starting to think how to develop new tests for AI. By definition, the tests would need to assess functions that humans are good at and computers aren't. Which brings us to the survey economy.

DIGITAL PIECEWORK AND THE SURVEY ECONOMY

Digital piecework in the information-gig economy looks a lot like the social science research model common on college campuses—if, that is, those research studies were being performed without ethical guidelines and at exploitative levels of compensation.

The survey economy is the landscape of underpaid labor in which people are paid pennies to complete online surveys, all of which require detailed personal information and most of which will in turn be used to create more detailed personal profiles of those individuals—making them more susceptible to microtargeted advertising. Underneath the high-technology veneer, the survey economy is the new manifestation of what used to be home-based sweatshop labor.

In nineteenth-century America, an entire underclass of women eked out a living sewing shirts for pennies apiece. Despite sixteen-hour days, they often earned only enough money to subsist, toiling for shop owners who found fault with the workmanship and refused to pay for their work.[5] Multiple layers of exploitation were involved as higher-paid male tailors outsourced labor to seamstresses who could be paid substantially (about 25 to 50 percent) less than men.[6] As the women's underpaid labor made it possible to produce clothes at lower cost, these pieceworkers provided the labor to mass-produce clothing for slaves in southern states. As technology advanced, sewing machines enabled productivity to leap. Yet, the workers didn't see more income; instead, they were encouraged to use installment plan payments to purchase the machines that would keep them toiling away for substandard wages.[7] Owners of capital commanded the resources to possess one group of humans as property and to extract poverty-level wages and ongoing debt from another group.

It may seem like a stretch to compare nineteenth-century sweatshops to today's freelance digital survey economy, but there are parallels in the business models. Dozens of websites like Survey Junkie, Vindale Research, LifePoints, and Global TestMarket invite users to spend hours exchanging personal information for pennies in service of the digital advertising ecosystem. The heart of their pitch is, "Take Surveys. Get PAID. Be an influencer."[8] The sites often post frothy "testimonials" with quotes that purport to be from real people, gushing that taking online surveys "provides an interesting way to make a little cash."

The marketing is deliberate. "Influencers" in the social media world are those Instagram-famous celebrities with millions of followers who are often famous for nothing more than being famous. No one in today's social media environment seriously believes that someone getting paid pennies for completing online surveys truly is an "influencer." But the pitch is intended to appeal to our sense of identity, to create the hope that we can have greater relevancy in the massively networked digital world where it's so easy to lose our true selves amidst the forest of idealized posts from friends, strangers, and influencers about all of the ways that they're crushing it in life. And to achieve all of these benefits, we just have to be willing to give up some time—and a great deal of our personal data.

As it happens, many of the people filling out Survey Junkie's online surveys *aren't* crushing it in life. Often, they're college students or working class and working poor—people with enough income to have a smartphone (which, like those nineteenth-century sewing machines, they purchased on an installment plan), but who are living in tight enough financial conditions that making even a little bit of extra cash is worth doing if there's a legal way to do it.

In reality, the pay is negligible. First, nearly all of the surveys begin with detailed demographic questions; these can take twenty minutes to answer, only for the user to discover that, because they don't meet the demographic criteria, they're disqualified from participating in the survey. Second, the highest-paying surveys require the user to activate their camera and allow the survey website to capture photos and/or video of the survey-taker. Third, sites typically pay on a points basis, and although they're eventually redeemable for gift cards or cash, they operate a bit like frequent-flyer programs: it takes a long time to accumulate points, there are lots of restrictions and exceptions, and sometimes the value expires before it can be used.

It helps to walk through specifics to put this in context. The typical conversion rate is one penny per point, or one hundred points to earn a dollar. Point allocation varies widely, from as much as forty points for a five-minute survey, or as little as forty points for a twenty-minute survey. If a user completes three surveys that pay forty points for twenty minutes, that user would earn the equivalent of $1.20 per hour. Even the more lucrative surveys don't pay much: if a user completes sixty of those surveys, each paying ten points per minute for an entire hour, they would still only earn $6 worth of points, for an hour's work—less than the federal minimum wage.[9] Even that $6/hour is likely unattainable; more often, the best a person can hope for is the chance to complete one to three surveys per week, for a total of about $12.[10] Compounding the low rate of return, many sites don't allow users to redeem their points until they've reached ten dollars (one thousand points), twenty dollars (two thousand points), or more. Some don't allow PayPal cash-outs, only gift cards or redemption by check sent through the mail. Many of them set expiration dates for points, requiring users to keep engaging with the site and providing more information in order to get paid for the work they've already done.

These sites have a number of features in common. They entice users by touting the advantages of free enrollment—an advertising mechanism that's long been known to channel consumer thinking into a narrow focus on only one dimension (free enrollment!) of a multi-sided decision. (Other questions should include: How much does it pay? How much of my time will be required? How much personal data do I have to give up? What will the site do with my data? Who else will my data be shared with? Could I more profitably spend the same time doing something else?) They offer users the opportunity to feel as though they're part of something larger than themselves—having a direct and personal role in shaping the direction of future products, advertisements, movies, television shows, and the like. Many of them emphasize a sense of community—even though interaction with other users isn't part of the online survey experience. And they're all heavily promoted on websites that brand themselves as offering moneymaking ideas to people who are "frugal," "cash-strapped," "savvy," looking for "life hacks," or other catchphrases that can be used to appeal to a person's idea of themselves.

The upshot? Slick advertising and an appeal to our sense of personhood are used to entice digital pieceworkers to accept pennies in exchange for hours of their time and detailed personal information. An earlier chapter of the book examined the question: What is privacy worth? In the online survey economy, just a few pennies. And you have to sacrifice hours of your time, along with your data, to get even that.

INSTITUTIONAL REVIEW BOARDS AND UNCONSENTING RESEARCH SUBJECTS

For nearly half a century, medical and social science research in the United States has followed a set of ethical guidelines under which Institutional Review Boards (IRBs) review experiments to make sure they meet ethical standards.[11] The National Institutes of Health maintains the regulations, which explain that research involves "systematic investigation designed to develop or contribute to generalizable knowledge," and defines human subjects research as research involving living individuals "about whom data . . . are obtained, used, studied or analyzed through interaction/intervention" by researchers who are using, analyzing, or generating "identifiable, private information."[12] The regulations set limits on the collection and use of "identifiable private information" from individuals.

In nearly all cases, IRBs require researchers to obtain informed consent from research subjects, explaining—among other things—what the research will entail, what the risks are, whether and how their information will be kept confidential, whether they are likely to experience any benefit or harm from participating in the research, and informing them that they can withdraw from the research at any time without penalty.[13] In deciding whether to approve research, IRBs look at a number of factors, including whether there are adequate provisions to protect subjects' privacy and maintain the confidentiality of their data, and whether the subjects are likely to be vulnerable to coercion or undue influence.[14]

Strictly speaking, the NIH regulations only apply to research that is supported, funded, or carried out by federal agencies or federally funded entities, but they're widely accepted and used across traditional research organizations. By every conceivable measure, the online survey economy fails to meet the ethical standards set in the NIH's human research subjects guidelines. The surveys are directed at people who, because of their financial situation, may be vulnerable to exploitation; the participants' personal data isn't kept confidential; the surveys don't disclose enough information about how participants' data is sold or used for survey takers to be able to give meaningful informed consent; the surveys' low rate of pay is arguably exploitative; and there's little if any benefit to the participants.

PERSONAL DATA, DIGITAL SWEATSHOPS, AND THE INTERSECTION OF PRIVACY AND PERSONHOOD

Taken together, what these digital phenomena illuminate is a massive market of unpaid labor (CAPTCHA), underpaid labor (online surveys), and informal (and sometimes unethical) social science research, all being conducted with no oversight, no ethical review, no meaningful form of consent, and all (except CAPTCHA, which falls equally on persons of all social rank and status who have internet access) in a fashion that takes advantage of those who are most desperate to snatch up leftover crumbs from a digital economy in which a new class of tech robber barons has made it possible to resurrect sweatshop labor and exploit our personal data.

It's striking, perhaps, that Turing was criminally prosecuted and stripped of his livelihood for personal behavior of the very type that many people hope to keep private today. Whatever one's definition of "privacy" might be, most people would conclude that privacy ought to include controlling information about one's sexual orientation or gender identity, their sexual behavior and intimate relationships. Many scholars also argue that privacy includes the ways that information from and about us bolsters, or diminishes, our sense of dignity and worth.

Computer algorithms may be getting closer to passing the Turing test. But it appears that, when it comes to the willingness to extract personal information from a new class of pieceworkers, the companies that benefit from having our data are behaving no more humanely than the capitalists of the Gilded Age did over a century ago.

CHAPTER 13
THE STALKER IN YOUR PHONE

Before smartphones, a jealous ex or an abusive partner might have listened in on your phone calls, followed you to or from work, or hired a private detective to watch the places you go and people you talk to.

Today's technology opens up a new world of opportunities for abusive partners to track the people—usually women—who are the objects of their obsession. These are often sold under the guise of location tracking services that can be used in lawful circumstances such as work-issued phones or a child's phone that's being monitored by a parent. But on the street—in chatrooms on the internet, YouTube videos, and Reddit threads—the software is known as "spouseware," more accurately reflecting what the real use case is. It's most often surreptitiously installed by adults to track their intimate partners. Sometimes, it goes to a further extreme, supporting government-sponsored repression and control of women. In both cases, the functions are largely the same: gathering information in order to exert control.

SPOUSEWARE AND STALKERWARE

Divorce courts and family law cases are filled with stories of jealous exes installing spyware on their partners' phones, either during the relationship to follow up on suspicions they might be cheating, or afterward, for ammunition to use in a contested or messy divorce case. In nearly all cases, installing spyware on someone else's phone is illegal. But it's also a booming industry, and there's an entire subterranean ecosystem on the internet that's designed to make it easy to do.

The statistics are alarming: according to the UK-based domestic violence prevention organization Women's Aid, nearly a third of all women who have experienced domestic violence have also discovered that their abusers installed GPS tracking services or other spyware on their mobile phones.[1] NPR has found that number to be even higher, with 85 percent of domestic violence shelters they surveyed in the United States reporting that they are working with victims who have experienced some form of online surveillance and stalking.[2] Seventy percent of the shelters said they have residents whose abusers eavesdropped on their telephone calls using remote activation of software that allowed them to listen in surreptitiously. Nearly three-quarters of the shelters place restrictions on their residents' ability to use social media platforms such as Facebook, because of the location tracking services associated with Facebook posts.[3] The US National Network to End Domestic Violence found that 71 percent of abusers use digital surveillance against their victims.[4] Similar numbers have been found in Australia and Canada, detailed in a recent report titled, "The Predator in Your Pocket," from the Citizen Lab, an interdisciplinary academic research component of the Munk School of Global Affairs and Public Policy at the University of Toronto in Canada. The Citizen Lab report paints a picture of widespread use of mobile phone surveillance tools as a means to carry out harassment, stalking, threats, and abuse of intimate partners and former partners online.[5]

Dozens of apps, mostly intended for Android phones, are already on the market, and new apps are popping up with alarming regularity. They advertise on Twitter and other social media platforms, with photos showing screen shots of a man lying in bed looking at notifications on his phone: "Helen just entered the nightclub" and "Helen left the office."[6]

Is this legal? Generally not. Installing software on someone else's device, or obtaining unauthorized access to their computing devices or accounts, is prohibited in the United States under the Computer Fraud and Abuse Act (CFAA). Other specific actions (like using spyware to eavesdrop on someone's conversation) are generally prohibited under the Federal Wiretap Act and analogous state laws. And when intentionally used against a spouse, significant other, or former partner for purposes of levying threats, instilling fear, planning to do harm, or finding their location in order to carry out physical violence against them, digital threats may come under the purview of laws banning stalking, threats, and violence.

Why, then, is "spouseware" or "stalkerware" such a lucrative business? The official advertising and marketing campaigns for these tools typically focus on other uses that would be lawful in most cases. These include anti-theft (being able to track your own phone if it's stolen), employer use (which is generally permissible if it's installed on an employer-provided phone and the employee has been notified that their use of the phone will be monitored), and for children and teenagers (whose mobile device use may be lawfully monitored by their parents or guardians). These official ads are often booby-trapped with code that shows lawful uses aren't the companies' entire marketing strategy. For example, the Citizen Lab found that the website of one spyware company included hidden HTML code that referred to spying on your spouse. That language wasn't openly displayed on the website but was visible to search engines, so that internet users running a search for how to spy on their spouse would find this company's software included among their search results.[7]

What exactly do these spyware apps do? Quite a lot. They allow the stalker to see their victim's real-time location data, of course, but that's just the tip of the digital iceberg. Many of them offer dashboard displays to allow the abuser to easily review and make sense of all the information available through the spyware: contacts, full recordings of phone calls, copies of text messages sent and received, photos, video files, a record of all their keystrokes in any application on the device (which can be used to obtain passwords and login information for other accounts), calendars, and a record of the phone user's internet history, including the websites they've visited and searches they've conducted online. Spyware can also enable abusers to carry out a number of other activities aimed at their victims, including breaking into and monitoring instant messaging accounts; breaking into email, social media, and other online accounts; impersonating their victim on email and social media; hacking into the victim's computer; nonconsensual distribution of intimate images and video (frequently known as "revenge porn"—more on this on page 166); and covert surveillance and surreptitious recording of the target through a hidden camera or webcam.[8]

Despite the fact that hacking into someone's phone is a criminal act in the United States and elsewhere, not many courts have addressed stalkerware, and there are few reported cases that would help illuminate how often its use is prosecuted. Laws like the CFAA allow victims to sue hackers for damages, but the amount of recovery is subject to a statutory cap. It isn't clear yet whether state or federal courts would allow additional compensation for emotional distress. In the divorce context, phone hacking often arises in connection with the admissibility of evidence (a spouse using photos from their partner's phone as evidence of an affair); it seldom arises as a stand-alone lawsuit—again, at least not in reported cases. In other words, while some legal remedies are theoretically available, pursuing them would often be costly and uncertain, and many victims either choose not to do so or, depending on their jurisdiction, are left without any meaningful recourse.

Spyware has become so widely used that the US Federal Trade Commission has issued guidance on how to figure out if an abusive ex or partner is stalking you, and what to do if you think they are.[9] Calls for more vigorous protections against spouseware and stalkerware have come from celebrities like Jennifer Lawrence who have found their nude photos leaked online; from elected representatives like Ron Wyden who are seeking to strengthen laws against GPS tracking of individuals; and from academics and legal experts such as Danielle Citron, a law professor at Boston University who has written extensively about the challenges of sexual privacy, intimate information, and spying on the internet, among other related problems.[10]

As if these privacy concerns weren't enough, journalists and researchers have also demonstrated the ways in which someone's roommate, intimate partner, family member, or friend can change the settings on perfectly ordinary apps, such as Google Maps and Find My Friends. They can track a mobile phone user throughout the day, or for days on end, without the phone's owner ever being aware that their location information is being sent to someone who changed their app settings without permission. It's much harder to design legal or technical protections against this kind of location monitoring, since it means the stalker has physical access to the phone and is likely in an ongoing relationship of some kind with the phone user.[11] Figuring out how to approach this kind of privacy problem may not be straightforward, but for the one in four women and one in nine men who report stalking or physical abuse by a current or former partner, finding solutions may be essential to both their safety and peace of mind.

GOVERNMENT-SPONSORED OPPRESSION IN THE HANDS OF INDIVIDUALS: THE SAUDI ARABIA "YES, SIR!" APP

The Kingdom of Saudi Arabia (KSA) has a long and well-documented history of enforcing laws that restrict the rights, privacy, and freedoms of its women. Under Saudi law, every woman must be under the custody and control of a male guardian, usually her father, husband, brother, uncle, or son. A woman may not marry or divorce without permission from her guardian; she may not travel unless her guardian allows her to; until 2008, she was prohibited from any kind of employment unless her guardian permitted it (those restrictions have been loosened somewhat in the decade since then); she can't receive medical care or be released from prison unless her guardian says he approves.[12] Throughout the twentieth century, women in Saudi Arabia were non-persons in the eyes of the state: they received no birth certificate when they were born. Although boys in Saudi Arabia were issued ID cards at the age of one year, no such card or registration was provided for girls. They

REVENGE PORN AND SEC. 230 OF THE COMMUNICATIONS DECENCY ACT

In 2017, a scandal rocked the US Marine Corps: a private Facebook group called Marines United was replete with posts of nude and intimate photos of women—including other service members, civilians, and current and former girlfriends and wives—along with graphic comments about assaulting and raping the women. Some 30,000 current and former Marines were members of the group, yet the content wasn't taken down until a new member spotted the apparently nonconsensual content and reported it to Facebook. The incident prompted congressional hearings, Facebook action, and soul-searching among Marine Corps leadership over how content could have accumulated that was so at odds with military discipline and the standards imposed under the Uniform Code of Military Justice. The incident served as just one example of the growing landscape of "revenge porn" available on the internet.

The simplest definition of revenge porn is the online posting of intimate photos or videos—nude, semi-nude, or sexually explicit—of individuals, usually women, who haven't consented to the distribution of the content. In some cases, it involves sharing nude photos or videos as part of a group text or online group chat. In other cases, it involves uploading of sex videos, either ones that were made consensually during a relationship or ones that were made surreptitiously, when the victim didn't know that a camera was on or while the victim was drunk.

Like so many other privacy issues, the explosion of revenge porn has been made possible by the rise of data-driven technologies. Platforms like YouTube made it easy to post user-generated content and prompted spin-off sites dedicated to user-generated porn. Smartphones made it possible to take, text, and upload sexually explicit photos—and with a few clicks made it possible to take the short step between consensual sexting and nonconsensual uploading of that same content to the internet. In the past decade, sites devoted to revenge porn have sprouted up all over the internet, many of them with names that seem designed to entice and titillate prospective visitors by implying that the content is of "real" ex-girlfriends, ex-wives, and other women whom the poster personally knows.[13] Alongside explicit photos and video, many of the sites also list the victim's true name and include links to her social media accounts. The result: the women often face public embarrassment or even humiliation in their relationships with employers, classmates, family, and friends, along with harassing comments from online strangers who see the posts as an opportunity to post or direct-message rape fantasies, assault threats, and more.[14]

Revenge porn has flourished in part because of the unintended effects of a law known as Sec. 230 of the Communications Decency Act, a law that functions as a case study of the ways in which technology-focused legislation can, over time, have an impact that is widely different from, or even the opposite of, what it was intended to achieve. First passed in 1996, the law's cosponsors, Senators Chris Cox and Ron Wyden, hoped that declaring online platforms were not to be considered "publishers" of information meant that the sites would voluntarily take on content moderation: that is, that they would carry out some form of screening for offensive or inappropriate content, or at least take it down when notified that the content had been posted on their sites. What the law did instead was to immunize online platforms from any liability within the United States for content that was posted by third parties on their sites. From negative online reviews to hate speech to revenge porn, platforms like Facebook, YouTube, 8Chan, Reddit, and more were completely free from any legal consequences for the content published by their members. In theory, a victim of revenge porn could take action directly against the person who posted the content. But only if the victim could find out who they were, and in many instances, the platforms allow users to create pseudonymous accounts and post information without attribution.[15]

Most Western democracies value free speech on the internet, but none of them grant platforms the wide-ranging scope of immunity that US law does. With hate speech, revenge porn, and political disinformation on the rise, Congress is considering proposals to modernize Sec. 230 of the CDA, potentially imposing some degree of accountability on platforms that knowingly host the most extreme forms of speech. It isn't clear yet what those reforms might look like, or how successful they might be.[16]

"DECLARATION OF THE INDEPENDENCE OF CYBERSPACE"

In 1996, after attending a party at the World Economic Forum in Davos, John Perry Barlow was incensed by what he saw as politicians' misguided attempts to regulate the internet; he was particularly irritated by President Bill Clinton's signature on Sec. 230 of the Communications Decency Act.[17] Barlow went back to his hotel room and whipped out an email that he sent to six hundred or so friends. In it, he declared that the internet was a space of its own, independent of any sovereign power's authority or ability to regulate. Barlow put forward a utopian vision of cyberspace as a realm of pure thought where participants would self-organize to bring about a just society based on a new vision of global interaction. Governments, he wrote, didn't "understand" internet culture. Furthermore, the internet's denizens hadn't invited governments to come play in their new sandbox; consequently, governments weren't welcome. Barlow wrote:

> You claim there are problems among us that you need to solve. You use this claim as an excuse to invade our precincts. Many of these problems don't exist. Where there are real conflicts, where there are wrongs, we will identify them and address them by our means. We are forming our own Social Contract. This governance will arise according to the conditions of our world, not yours. Our world is different.[18]

Barlow's vision was naively independent of a host of realities: the fact that the electrons that travel through cyberspace take their paths across physical infrastructure, owned by a combination of corporations and nation-states; the fact that behavior on the internet is shaped as much by norms as by laws, and both norms and laws vary widely across the global cultures connected by the physical infrastructure of the internet; and the fact that what happens in cyberspace reverberates through the physical world.

Indeed, in the years since Barlow published his screed, countless examples have demonstrated the intersection between online behavior and the physical world. When online gamers engage in "swatting"—calling police with a false report of a domestic disturbance or other threat as revenge for an online slight—real people get hurt, arrested, or even killed.[19] When internet users dox their targets—publicly posting information about the address and other contact details of someone who they dislike or disagree with—the doxing targets find themselves harassed, threatened, and sometimes hounded out of their homes.[20] When bitter exes post revenge porn or when fans or pranksters post pornographic deep fakes, the emotional and reputational damage to the victim has tangible effects in the physical world; the personal damage is real.[21]

In Barlow's vision, all would be equal on the internet. In his manifesto, he wrote that, "We are creating a world that all may enter without privilege or prejudice accorded by race, economic power, military force, or station of birth."[22] On the twentieth anniversary of his declaration, Barlow was quoted as standing by it, believing—despite criticism from a host of commentators—that his manifesto has stood the test of time.[23]

However, the internet has often served as a platform for reinforcing the inequalities and power imbalances that exist in the physical world. Women are disproportionately targeted by revenge porn. Some of the most famous episodes of doxing—from the trolls who published private information of gaming community critic Zoë Quinn in the controversy that became known as Gamergate[24] to the repeated publication of Christine Blasey Ford's home address following her testimony against Supreme Court nominee Brett Kavanaugh[25]—have targeted women. Further, women are often the targets of heightened spite and vitriol on social media and other platforms.

Contrary to the hopes of early internet enthusiasts, the evidence suggests that cyberspace and the physical realm are in fact inextricably linked, and that the equality Barlow and others had hoped for has not come to pass. On the contrary, the internet has made it possible for our digital lives to be used against us in ways that carry tangible, negative, real-world impact, and that reinforce differences in privilege and status: we can't escape our bodies or physical surroundings simply by interacting in cyberspace.

couldn't receive a passport in their own name. A woman's very identity existed only as an extension of a man: they had no way to prove who they were—and if they went missing, there was no way to prove, over the objections of a male guardian, that such a woman had ever existed. If a woman wanted to testify in court on any kind of matter, first she had to find two men who were willing to go to court with her to testify as to her identity—and that was just a first step before finding four men who would testify as to the accuracy of whatever statements or complaints she might make.

Although the standards of women's rights in KSA are still drastically below those afforded by nearly every other nation in the world,[26] there has been some incremental progress. In 2001, the government began issuing identity cards for a small number of women who had their male guardian's permission; the ID cards were not issued to the women—they were issued to the guardian.[27] In 2006, women were allowed to obtain identity cards without permission from their male guardian, and in 2013 identity cards became mandatory for women, just as they had been for men since the last century.[28] In 2016, the Saudi government for the first time allowed women to receive a copy of their own marriage contract—a right that had previously been reserved solely for men.[29] Theoretically, at least, being in possession of a copy of the marriage contract could pave the way for women to enforce a limited range of rights against their husbands in the future. And in 2018, women in Saudi Arabia were finally, after years of protests, imprisonments, and petitioning, granted the right to drive.

What does this have to do with personal data? Saudi women have faced a unique set of challenges: living under a government that denied central dimensions of their personhood by placing nearly all decision-making autonomy into the hands of male guardians, and prohibiting any form of legal identity unless that identity was attested to by men. At the same time, however, this lack of identity didn't protect Saudi women from surveillance or intrusion by the state: Saudi women protesters were surveilled, arrested, and—according to some reports—interrogated, beaten, and tortured. The Saudi government has extensive internal surveillance capabilities directed toward dissent of all kinds, and that includes monitoring of social media and other electronic communications.[30] As a result, Saudi women were sure to rise to the attention of government authorities, even if they themselves could not benefit from that attention by exercising the rights and privileges that men have in Saudi society.

Against this backdrop, in 2011, the Saudi government unveiled a new website that would allow male Saudi citizens to carry out a range of ordinary government and business transactions online, rather than having to go in person to a government office building and stand in line. In 2015, the Saudi government released a mobile app, titled Absher, which roughly translates to "Yes, sir!" The app, which was released on the Apple App Store and Google Play,[31] makes it possible for users to carry out those web-enabled interactions with government offices from their personal mobile platforms.

The app doesn't provide real-time location tracking. However, because Absher connects to government functions like travel approvals and passport permissions, it's used by male guardians to track women's travel and interact with government services indicating whether the women are, or aren't, permitted to travel or work. The app's potential to be used as a means of denying women the right to travel has prompted criticism of Google and Apple for making it so widely available on their platforms.

Despite all of the criticism, other voices, including some women in Saudi Arabia, have expressed their support for the app, saying that while it supports bureaucratic functions that reinforce male control over women's autonomy, the app has provided an overall benefit to women by making it easier for male guardians to grant permission for things like travel and work. In other words, for those commentators, the app isn't the problem; the law is. And while the law may be repressive, at least the app helps facilitate a streamlined ability for women to exercise the limited rights that they have.

Examples such as this raise important questions about whether and to what extent multinational corporations like Apple and Google should be expected to monitor or restrict the types of apps made available on their platforms, and be held accountable for the ways in which their technologies are used. Multinational companies often argue that they face complex challenges in complying with widely varying legal regimes in an international market. Whether the criticism is of Google for creating a China-compliant search interface that supports Chinese government censorship, or of Apple for removing essential privacy protections for phones being marketed in China, there's an important policy debate to be had about whether companies should maintain a value-neutral stance on the privacy implications when they knowingly earn profits from products that are used or adapted to support intrusive collection of personal information and related abuses, whether those abuses are tied to discrimination based on gender, religion, or sexual orientation, religious or political oppression, or other concerns about fundamental human rights. An intrinsic part of the debate should be: Where companies have been able to flourish because they were founded in, and remain headquartered in, the United States and other Western democracies, is it appropriate to impose an ethical or legal obligation for them to ensure that their continued profits don't take advantage of repression elsewhere?

SECTION IV

Who's Your Big Brother?

I was sitting in a conference room at the Thompson Hotel in Chicago in June 2018 for a talk I was giving called "The Dangers of Data." A mildly alarmist title, perhaps, but a fitting one nonetheless for this group of trial lawyers who were trying to understand the ways that ubiquitous data collection was shaping legal risk for their clients. Midway through the discussion, a woman in the audience—a data-savvy lawyer with plenty of cybersecurity experience—asked, "So at CIA and NSA, do people just walk down the halls chuckling to themselves and rubbing their hands with glee, because there's so much information about everyone everywhere? Do they feel like their job's never been easier?"

I shook my head.

"Just the opposite," I answered. "In today's world, in democratic countries like ours, the government doesn't have nearly as much data as private companies do. I suspect some folks in the Intelligence Community wish they could access more of it—it's got to be galling sometimes, knowing that Google and Facebook have so much more information than they do." I paused. "Of course, it's different in countries like Russia and China. Authoritarian countries have a level of data collection on individuals that really is staggering. Just look at China's network of surveillance cameras in cities, and the program that has individual citizens giving each other social credibility rankings. But no, we don't do that here."

I sensed that some of the attendees were skeptical about my response. But after working inside the IC, and on Capitol Hill overseeing it, I can say this really is true: the US government has a great deal of data about individuals, but far less than many people think. In most federal agencies and for most kinds of data, there are strict controls on what can be collected, how long it's held, and how it gets used. And—despite what cynics might think and civil libertarians might fear—failures to abide by those rules get discovered and the corrective actions are usually much harsher than a slap on the wrist.

But there are new technologies emerging every day, creating new types of data and more of it, and expanding ways to collect and use that data. There's no question that the laws and regulations for government use of data are struggling to keep up, and some government agencies are lagging behind when it comes to developing and implementing rigorous policies and over-sight to govern their use of personal data. The mismatch between policy and technology creates thorny new questions and opportunities for abuse.

And that's just in this country, where we have a long-standing tradition of transparency and oversight. Critics often argue that US surveillance activities should be more transparent. But the truth is that, from inspectors general to the Foreign Intelligence Surveillance court to civil litigation and other means, we have one of the most carefully scrutinized, intensively overseen, and remedy-rich approaches to government surveillance of any nation on earth. The United States is by no means the only country carrying out eaves-dropping and data analysis. We're just more open and public about what we do and how we do it. Few nations share the full mix of attributes intrinsic to the United States, where skepticism toward government is embedded in our cultural DNA, where our laws embrace a near-total commitment to free speech and press, and where our three-branch system of government provides an exceptionally robust opportunity for checks, balances, and oversight when the interaction among those branches is working in the manner in which our constitutional framework was designed. The fact is that nearly every nation on earth with a significant GDP has intelligence and surveillance programs similar to ours. Just how advanced and comprehensive those programs are is often more limited by resources and technical know-how than by policy, ideology, or law. Some of the nations that make the loudest proclamations about the sanctity of privacy as a human right are the same nations whose intelligence activities require no independent judicial review, receive no regular parliamentary oversight, and have no other institutional tradition of independent watchdogs.

Unlike much of the data gathering that happens in the private sector (discussed earlier in the book), most government surveillance in the United States is limited by an established legal framework that starts with the Constitution and has evolved through centuries of case law. The framework isn't perfect: reasonable people frequently disagree on what boundaries should be captured in the law, and both case law and statutes struggle to keep pace with technological change. Nonetheless, a framework exists, and it's a powerful and important one. This section briefly describes the history of that framework and how it's evolved in recent years to try to catch up with the challenges of a digital world. It looks at some of the areas where privacy rights are most at risk from federal activities, including border searches in the United States. This section then tries to identify a balance between well-bounded programs with effective controls and those that are at risk of running amok. In other words, it makes the case for why we still need government surveillance in the era of Trump, and how we can protect ourselves from its overreach.

CHAPTER 14
THE US INTELLIGENCE COMMUNITY POST-WWII

Just Because You're Paranoid Doesn't
Mean They're Not Watching You

Most of the agencies of the US Intelligence Community (IC) have their roots in the years immediately before and after World War II. During those years, the national security threat from adversarial foreign governments was pressing and apparent, and the agencies charged with understanding foreign adversary plans and intentions were granted broad authorities with few restrictions and little oversight. This lack of structure and accountability came to a head in the 1970s, in a series of congressional hearings that led to widespread and badly needed intelligence reform. That overhaul served the country well for decades, with robust oversight in place and—although there were notable, important exceptions—there were few abuses of the power that the IC was granted.

Following the terrorist attacks of 9/11, the nation sensed that it was once again under attack by adversaries, and this time, the threat might not be coming from adversarial nations, whose primary actors could be easily identified, but from international terrorist organizations, whose leadership was hard to identify, whose plans and intentions could be harder to discern, and whose operatives could easily hide within the US populace. In response to this shifting threat, the IC began expanding its efforts in ways that raised echoes of the concerns of the past: Did the techniques being adopted by the IC after 9/11 create risks of the same kinds of government overreach that prompted the intel reforms of the 1970s? Would a new set of reforms be needed to address the risks of national security surveillance in a post-9/11, data-driven world? In order to understand how we can prevent and detect government surveillance overreach, and curtail it when it happens, it's important to know what our history with those issues is and to figure out, while testing the limits in today's technology environment, where past solutions can succeed again.

THE GROUNDWORK FOR US AND INTERNATIONAL DATA PRIVACY MODELS

In 1973, the US government was expanding its use of computer technology to manage and store information. Even in those early days, it was possible to imagine countless ways in which a computer's ability to store, index, search, and retrieve information could be used for mischief or to inflict outright harm. Against that backdrop, then US Secretary of Health, Education, and Welfare (HEW) Elliot L. Richardson commissioned an advisory committee to conduct a study on "Records, Computers, and the Rights of Citizens." The committee's charge was to analyze and make recommendations about the "harmful consequences that may result from using automated personal data systems," as well as potential safeguards to keep computers from causing harm, and mechanisms to redress harm when it occurred. (The committee also looked specifically at policies relating to the use of Social Security numbers—an issue so narrow in the landscape of today's personal data challenges that it almost seems quaint.)[1]

The advisory committee issued its report in 1973, acknowledging both the benefits and dangers of computerized holdings of personal data. Caspar Weinberger, who succeeded Richardson as HEW Secretary, endorsed the group's work, writing a foreword noting the importance of striking a fair balance:

Computers linked together through high-speed telecommunications networks are destined to become the principal medium for making, storing, and using records about people. Innovations now being discussed throughout government and private industry recognize that the computer-based record keeping system, if properly used, can be a powerful management tool . . .

Nonetheless, it is important to be aware, as we embrace this new technology, that the computer, like the automobile, the skyscraper, and the jet airplane, may have some consequences for American society that we would prefer not to have thrust upon us without warning. Not the least of these is the danger that some record keeping applications of computers will appear in retrospect to have been oversimplified solutions to complex problems, and that their victims will be some of our most disadvantaged citizens.[2]

The committee's recommendations would lay the groundwork for all of the US and international data privacy models to follow. In its report, the committee recommended the adoption of a "Code of Fair Information Practice"—a set of guidelines that later became known as the Fair Information Practice Principles (FIPPs) or Fair Information Practices (FIPs).[3] According to the committee, the federal government should incorporate the FIPPs into all of its computer systems, with the result that:

- There must be no personal data record-keeping systems whose very existence is secret.
- There must be a way for an individual to find out what information about him is in a record and how it is used.
- There must be a way for an individual to prevent information about him that was obtained for one purpose from being used or made available for other purposes without his consent.
- There must be a way for an individual to correct or amend a record of identifiable information about him.
- Any organization creating, maintaining, using, or disseminating records of identifiable personal data must assure the reliability

of the data for its intended use and must take precautions to prevent misuse of the data.

The report was careful to note that the "safeguard requirements" set out in the FIPPs "define *minimum standards* of fair information practice."[4] In other words, these privacy principles should be viewed as a floor, not a ceiling, for privacy best practice.

The FIPPs were quickly recognized as commonsense baseline obligations for federal government collection of data. However, there are a number of exceptions, particularly in the law enforcement and national security contexts. And although Congress has passed sector-specific laws such as HIPAA for the protection of health-care information, the United States has no general-purpose data privacy law at the federal level that requires any private-sector person or entity to comply with anything approximating the FIPPs.

THE CHURCH AND PIKE COMMITTEE HEARINGS

The early seventies brought in a wave of political and social turbulence. The civil rights movement, opposition to the Vietnam War, and support for women's rights were deeply divisive cultural issues. Protesters marched by the hundreds of thousands in Washington, DC, and in smaller numbers in cities and towns across the nation. The heavy-handed response to these events by federal, state, and local governments served to deepen the existing mistrust that protesters had in America's institutions. News reports surfaced about murky CIA assassination plots. The Defense Intelligence Agency (DIA) was accused of sending undercover military intelligence officers to infiltrate domestic activist groups. The FBI was suspected of planting informants in civil rights and antiwar groups and spying on prominent Americans, from Dr. Martin Luther King Jr. to Dr. Benjamin Spock. The Watergate break-in of 1972, the Watergate hearings of 1973, and President Nixon's resignation in 1974 had all paved the way for a deepening mistrust of the executive branch of government.

In 1973, the Watergate hearings had revealed the CIA's involvement in what were clearly domestic political activities: Nixon's orders to have the CIA wiretap and steal valuable campaign-related information from the Democratic party establishment in hopes that this dirty-tricks approach would give him an edge in the upcoming election. Although the primary focus of the Watergate inquiry was on the actions of the president, the revelation that CIA officers had participated in the Watergate break-in and cover-up was a staggering admission of partisan activism that ran directly contrary to the important principle that US foreign intelligence activities should always remain nonpartisan in nature, focused on external adversaries rather than internal political horse races.

Not long after those revelations, in December 1974, the *New York Times* reported that the CIA had been spying on antiwar activists in the United States for years.[5] The article's lede dropped a bombshell:

The Central Intelligence Agency, directly violating its charter, conducted a massive, illegal domestic intelligence operation during the
Nixon Administration against the antiwar movement and other dissident groups in the United States, according to well placed Government sources.[6]

The CIA reportedly maintained files on at least ten thousand Americans and had, throughout the 1960s and early 1970s, carried out routine surveillance. It had followed and photographed demonstrators at protests, attempted to infiltrate antiwar protest groups with undercover CIA officers, and even surveilled an antiwar US congressman.[7]

At the time, there was a deep-rooted belief in some parts of the national security establishment that US domestic unrest—and especially the civil rights and antiwar movements—was being fomented by foreign governments. It wasn't a crazy theory: the Soviet Union in particular had long engaged in information operations to undermine the stability of their adversaries' internal political environments. Causing chaos around the world—and particularly in the United States and the West—served important domestic and foreign policy goals of the Soviet state. Turmoil in the United States allowed the Soviet government to argue to its own people that the West was not, in fact, a better place to live than the Eastern bloc. The Vietnam War was cast as capitalist colonialism, inherently abhorrent to communist ideals. And bolstering the civil rights movement not only gave the USSR an opportunity to drive Americans to hardened positions in opposition to each other, it also allowed the Soviet government to point to racial inequality in the United States as an example of how far short the American experiment had fallen when compared with its ideals.

For US government officials, understanding the distinction between Soviet information operations and homegrown protests could be a meaningful difference. Committed US protesters working to bring about an improved future in their home country might focus on paths to constructive solutions: voting in new elected officials, passing new legislation, filing impact litigation in courthouses, and pursuing dialogue with other Americans in hopes of changing their views. Foreign government agitators, however, had a primary goal of increasing chaos, disruption, and disillusionment in American life; they might attend the same protests, but their goals were far more likely to consist of stoking anger, getting opposing groups to harden their stances and see their opponents as enemies, and obstructing any reforms that could lead to meaningful progress. For these outside agitators, the goal wasn't to move the country forward, it was to throw kerosene onto the coals and see if they would erupt into open flames. Viewed in that light, it wasn't unreasonable for US officials to want to know what role foreign governments were playing in covert efforts to shape the social and political discourse of a deeply divided nation.

In the kernel of legitimate national security questions, however, lay the seeds of significant government overreach. Whatever the actual scope of the counterintelligence threat might have been, the Nixon White House had become obsessed with domestic protesters. In 1970, the administration called together the heads of the FBI, CIA, and the military intelligence agencies and charged them with gathering information on domestic dissenters and determining whether they were subject to foreign influence.[8]

Together, the CIA, NSA, DIA, and FBI developed the forty-three-page report that became known as the "Huston plan," named for Tom Huston, the White House lawyer and speechwriter who directed the work. In it, the agencies identified three categories of domestic threats: "militant new left groups" like student protesters, antiwar activists, and "new left terrorist groups"; the "black extremist movement," including groups like the Black Panther Party; and a catch-all of "other revolutionary groups" ranging from the Communist Party to the Socialist Workers Party and groups advocating for Puerto Rican independence.[9] The report explained that the Soviet intelligence threat was real. The US intelligence community assessed that the KGB, the Soviet Union's main security agency from 1954 to 1991, had some 300,000 personnel, with about 10,000 of those engaged in operations outside the USSR. The KGB was known to be engaged in both traditional espionage activities as well as "direct intervention in fomenting and/or influencing domestic unrest." As the report pointed out, "Taken in complete context, these services constitute a grave threat to the internal security of the United States because of their size, capabilities, widespread spheres of influence, and targeting of the United States as 'enemy number one.'"[10] Notably, the report didn't conclude that there *was* widespread or significant foreign influence—only that it was possible such influence existed.[11]

Faced with the possibility of foreign influence and the president's direction, the agencies' report presented the White House with options ranging from inaction to aggressive countermeasures. Huston, serving as the lead, argued in favor of the most aggressive approaches, which included allowing intelligence agencies to intercept and read the mail of suspected intelligence targets[12] and breaking into their homes and workplaces to gather evidence;[13] using undercover military personnel to infiltrate protest groups and recruiting "campus sources" (i.e., students) as government informants;[14] expanding electronic surveillance generally; and granting the NSA explicit authority to conduct electronic surveillance against US citizens for purposes of investigating and countering domestic unrest.[15] Huston's report acknowledged that some of these activities fell within gray areas where existing law or policy would have to be reinterpreted. Others, he acknowledged, were "clearly illegal." Break-ins, for example, "[amount] to burglary . . . It is also highly risky and could result in great embarrassment if exposed. However, it is also the most fruitful tool and can produce the type of intelligence which cannot be obtained by any other means."[16]

In June 1970, President Nixon signed off on Huston's recommendations.[17]

When the Watergate hearings brought the Huston plan for domestic surveillance to light, Congress recognized that US intelligence activities had gone unscrutinized for too long.[18]

In response to these snowballing allegations, in January 1975, a newly re-constituted Senate with a Democratic majority passed Senate Resolution 21, establishing a committee that would be formally granted the jurisdiction to oversee federal intelligence operations and determine "the extent, if any, to which illegal, improper, or unethical activities were engaged in by any agency of the Federal Government."[19] The predecessor of today's Senate Select Committee on Intelligence, this special committee was most often referred to as the Church Committee, after the man chosen to lead it, Idaho senator Frank Church. Not to be outdone, the US House of Representatives established a similar inquiry the following year, which would become the predecessor of today's House Permanent Select Committee on Intelligence (HPSCI). At the time, it was more colloquially known as the Pike Committee, after its chairman, Representative Otis Pike of New York.

The Church and Pike Committees launched a series of investigations that would transform the US Intelligence Community, resulting in enduring changes that lasted well into the next century. But before those things could happen, first the committees had to hold a series of very public, closely followed, hotly debated, high-profile hearings.

As it happened, thanks to a series of bureaucratic snafus, implementation of the Huston plan was delayed and eventually fell by the wayside.[20] However, according to Senator Church's opening statement in a congressional hearing held some five years later, the Huston plan really amounted to a continuation of a long saga of Intelligence Community overreach and disregard of the civil liberties of Americans. The Church Committee hearing transcripts and exhibits provide considerable support for that claim.[21]

As would be the case in the intelligence oversight committees of every future session of Congress, many of the committees' hearings were held in closed session due to the classified nature of much of their review. However, in the fall of 1975, a series of lengthy public hearings was conducted as well. The Church Committee held 126 full committee meetings and 40 subcommittee hearings, interviewed approximately 800 witnesses (some in public and many in closed sessions), and reviewed 110,000 documents. Its final report, containing 96 specific recommendations, was published on April 29, 1976.[22]

The Church Committee was conscious of the historic nature of its work, noting at the outset of its report that this was the first comprehensive review of US intelligence activities to take place since World War II.[23] The committee took pains to balance the nation's legitimate needs for intelligence gathering with the vital interests of liberty.[24] But it minced no words in its view that the national intelligence agencies had gone too far, engaging in unlawful behavior, and that a necessary part of the correction to that was robust congressional oversight—in part because of the temptation of well-meaning individuals to lose sight, in times of turmoil, threat, or strife, of the framework of laws that ought to guide them. In their final report, the Church Committee stated:

> The Committee is of the view that many of the unlawful actions taken by officials of the intelligence agencies were rationalized as their public duty. It was necessary for the Committee to understand how the pursuit of the public good could have the opposite effect. As Justice Brandeis observed:
>
> > *Experience should teach us to be most on our guard to protect liberty when the Government's purposes are beneficent. Men born to freedom are naturally alert to repel invasion of their liberty by evil-minded rulers. The greatest dangers to liberty lurk in insidious encroachment by men of zeal, well-meaning but without understanding.*[25]

Church's committee further noted that "the root cause of the excesses which our record amply demonstrates has been failure to apply the wisdom of the constitutional system of checks and balances to intelligence activities."[26] In assessing its task, the Church Committee wrote that the question it faced was in striking the same balance that would become a refrain in future decades: determining how the civil liberties of the people could be maintained while government attempted to protect their security. Faced with postwar threats, the committee wrote:

Too often intelligence has lost this [foreign] focus and domestic intelligence activities have invaded individual privacy and violated the rights of lawful assembly and political expression. Unless new and tighter controls are established by legislation, domestic intelligence activities threaten to undermine our democratic society and fundamentally alter its nature.[27]

The Church Committee remarked on past abuses of governmental power, from the Alien and Sedition Acts to the internment of Japanese Americans in World War II, but warned that the inherently secret nature of intelligence activities made them different, and perhaps even more dangerous. Other kinds of abuses of power were generally tied to statutes or executive orders that were publicly known. Intelligence activity, however, by its nature, is covert, sometimes carried out under classified orders and in such a way that its targets often never know what happened.[28] Like a petri dish, that secrecy created an environment in which what started out as limited intrusions on rights festered and grew into programs that far exceeded their original purpose and were inconsistent with the values of the nation.[29] The Church Committee noted:

Too many people have been spied upon by too many Government agencies and too much information has been collected. The Government has often undertaken the secret surveillance of citizens on the basis of their political beliefs, even when those beliefs posed no threat of violence or illegal acts on behalf of a hostile foreign power.[30]

The committee detailed the government's use of wiretaps, microphone "bugs," and surreptitious mail opening that "swept in vast amounts of information about the personal lives, views, and associations of American citizens." It described the government's decision to target groups based on their political views or lifestyles, or because they were deemed "potentially dangerous," even though they hadn't engaged in any unlawful activity.

The authors of the report laid the blame at the feet of all three branches of government: each had failed in some part of its duty to secure liberty along with security. The executive branch didn't delineate the scope of permissible activities or implement procedures to keep the IC acting within those guidelines. Congress had continued to appropriate money for intelligence activities without overseeing how the money was being spent. When domestic intelligence activities were challenged in court, the judiciary tended to be overly deferential.[31]

As a result, the US IC had investigated a "vast number" of American citizens and domestic organizations. The CIA and FBI had opened nearly 400,000 first-class letters, and the NSA intercepted millions of telegrams. FBI headquarters had more than half a million files on Americans, with countless more in FBI field offices around the country, and a list of some 26,000 people to be rounded up in the event of a national emergency.[32] The IC defined its targets too broadly and held the information for too long, collecting data on "proponents of racial causes and women's rights, outspoken apostles of nonviolence and racial harmony, establishment politicians, religious groups, and advocates of new lifestyles." The IC monitored protests of welfare mothers in Milwaukee, infiltrated church youth groups in Colorado, and sent agents to conferences where priests discussed birth control and a Halloween party for an elementary school where a local "dissident" might be present.[33]

Perhaps most concerning of all, none of these were isolated incidents. According to the Church Committee's report, under President Roosevelt and every other President since that time, the intelligence apparatus of the federal government had been used for domestic partisan political purposes by every White House, by both parties, through countless changes in foreign policy, domestic politics, and cultural developments within the United States.[34]

The government's improper use of intelligence capabilities imposed significant costs on individuals and on society. Efforts to discredit individuals led to the breakup of their marriages, damage to their friendships, and loss of standing at work. Media manipulation by the FBI—leaking information and planting stories in friendly news outlets—undermined political and financial support for leaders and causes such as Dr. Martin Luther King Jr. and the Poor People's Campaign. Deliberate intimidation of protesters chilled Americans' First Amendment rights to freedom of speech and of association. FBI's targeting of speakers, teachers, writers, and publications suspected of being involved in left-leaning causes inhibited the free exchange of ideas. And countless taxpayer dollars were squandered through expensive, long-running, intrusive, and widespread intelligence programs, some of which violated Americans' rights.[35]

Despite these many, now-documented excesses, abuses, and failings by the Intelligence Community, the Church Committee also underscored the continued need for legitimate intelligence activity in order to assess the plans and intentions of adversarial foreign governments and to ferret out international terrorism cells and other transnational threats to national security. As a result, the committee noted, the answer was not to do away with intelligence agencies or intelligence activities. Instead, the goal should be to properly bound them, to make sure that previously vague authorities were spelled out in clear detail and that agencies that might resist oversight or be reluctant to acknowledge their programs would have to face scrutiny from empowered and dedicated overseers. As the committee wrote in its report, "Clear legal standards and effective oversight and controls are necessary to ensure that domestic intelligence activity does not itself undermine the democratic system it is intended to protect."[36]

REFORMS THAT RESULTED FROM CHURCH AND PIKE

As a result of these investigations, sweeping reforms were put into place. Permanent select committees for intelligence oversight were established in both the House and the Senate. President Reagan issued an executive order that updated the roles and responsibilities of each component of the Intelligence Community, clarifying what the limits of their authorities were. The new presidential direction, which would become Executive Order 12333, required that each component of the IC have procedures, reviewed and approved by the attorney general, to minimize the collection, processing, retention, and dissemination of information about US persons. Congress passed the Foreign

Intelligence Surveillance Act (FISA) in order to regulate and restrict the use of electronic surveillance techniques in gathering foreign intelligence. Other reforms continued in the succeeding years: Inspectors general were appointed at the various IC agencies, as well as at parent components like the Departments of State and Defense. Internal intelligence oversight mechanisms were established within the IC to educate the workforce on its privacy and civil liberties obligations and train the workforce on how to implement programs to protect those important rights. Oversight officers within the IC reviewed programs, reported compliance violations, and developed the underpinning policies and procedures that would be necessary to support well-disciplined intelligence activities. The Department of Justice became more directly involved in overseeing intelligence activities, particularly those that required FISA Court permission. And independent boards and commissions were established, such as the President's Intelligence Advisory Board and, in later years, the President's Civil Liberties and Oversight Board.

Each of these measures made real and lasting differences in the ways in which the Intelligence Community approached its exercise of authority and the execution of its mission.

When the Church Committee wrote its report, phrases like "data privacy" hadn't yet come into currency. However, the committee's work focused on precisely the types of surveillance that, from the 1970s through today, have enabled governments to gather detailed information about individuals.[37] Today's concerns about government surveillance are, in many respects, questions about the ways that twenty-first-century data and digital technologies could heighten the impact of what are, fundamentally, twentieth-century models of policing and security. These lessons from oversight are enduring ones, both for government activities and for the growing concerns about private-sector data collection. A bipartisan, bicameral congressionally appointed review of private-sector data practices could learn a great deal from the Church and Pike Committees in how to carry out an effective investigation, what kinds of reforms might be recommended, and how to incorporate lessons from intelligence oversight into the regulation of private-sector data.

LESSONS FROM INTELLIGENCE OVERSIGHT

Despite our current sense that technological advancements have left us as a society, and as a global community, ill-equipped to deal with the challenges of balancing individual rights with national security, these challenges aren't as new as we might think. In 1976, the Pike Committee draft report noted:

> In the last half-century, electronic technology has revolutionized the science of investigations. These developments also mean that "Big Brother" may be watching.
>
> Improper application of electronic surveillance poses obvious risks because of its enormous potential for invading privacy and the difficulty of detecting intrusion.[38]

The fact that this sense of peril at the hands of technology has persisted across parts of two centuries shows that this is a problem that—despite changes in the details over time—is enduring. The changing shape of the law has demonstrated the need for it to keep pace with changing technology. And the continued reauthorization of FISA and related laws, along with the accumulated weight of countless court orders and executive branch actions, indicate that members of both major parties and all three branches of government have, for the past fifty years, recognized it is both possible and necessary to strike a balance between national security and individual privacy.

The Pike Committee made a number of recommendations in its draft report, and several members of the committee also put forward independent recommendations of their own. California representative Ronald V. Dellums wrote:

> These recommendations should stimulate extremely important and timely discussion, debate and consensus about such vital and basic questions as:

1. Is secrecy compatible with principles of democracy ostensibly embodied in our constitutional form of government?
2. If and where is secrecy necessary?
3. How much secrecy is necessary, and what forms should it take?
4. What safeguards against abuse are required?
5. What, if any, are our legitimate and necessary intelligence needs?
6. How much change, restructuring, and/or elimination of organizations are required to meet on the one hand the *legitimate* intelligence needs of our Nation, and on the other hand safeguard against abuse of people, power, and the Constitution?
7. As our world continues its rapid changes and shifts, what level of our already limited resources do we perceive as necessary to meet our intelligence needs?[39]

Dellums wasn't wrong to point to the risks of unregulated intelligence gathering and covert action by the Intelligence Community. And we wouldn't be wrong to apply many of those same principles and concerns to the widespread collection of personal data, the growth of artificial intelligence predictions based on that data, and the use of those conclusions to try to manipulate public opinion and individual behavior today.

CHAPTER 15

WHERE DO YOU DRAW THE LINE?

Data Collection in the US Intelligence Community Post-9/11

ollowing the terrorist attacks of September 11, 2001, the United States and other Western intelligence agencies went on high alert. There was a heightened sense of imminent threat, of the need to push intelligence capabilities to the very edge of what was permissible. Where previously some agencies had developed a post-Church-and-Pike cultural tendency to steer far clear of the dividing line between legal and illegal activities—taking great pains to ensure their actions were unquestionably on the right side of that line—now it felt as though there was a new imperative. This renewed assertiveness by the US Intelligence Community led to new laws as well as new programs whose legal authorization was less clear. From collection of bulk metadata to content collection programs like the one that later became congressionally authorized and known as "702 collection," Western governments were leaning forward on intelligence collection in ways that their citizens didn't always realize or understand. At the same time, the revolution in computing power that was affecting every other area of life was also impacting the US IC. From cloud computing storage to big data analytics, the

IC had to figure out how to deal with the ever-rising tide of digital data, and how to identify and divert the small streams of useful information that were coursing through that flood.

After September 11, the US Intelligence Community was faced with a series of urgent questions: Who were the attackers? Could the attacks have been foreseen? What could the US IC do to reduce the chances that the United States might be caught off-guard by another catastrophic event in the upcoming months and years?

Those questions would have been asked under any circumstances. But 2001 also marked the beginning of an era in which telecommunications infrastructure and technology were changing, and the global community was on the brink of an unprecedented degree of interconnection and interdependency. This new technological future had been heralded in part by the "Y2K" phenomenon, the scare that had generated global anxiety in the final months of the twentieth century as technologists, politicians, businesspeople, and pundits recognized that nearly every facet of modern life had become reliant to some degree on computer systems for effective functioning. With that realization came another, more sobering, thought: many of the older computer systems—the kind embedded in critical infrastructure components like power plants, electric grids, and telephone switching systems, as well as in everyday business application—were hard-coded with a two-digit field for dates, meaning that when the dates flipped over from "99" to "00," it was anyone's guess what would happen next.

As it happened, Y2K was largely a nonevent. The predicted catastrophes—air traffic control systems failing, widespread power outages, global commerce grinding to a halt—didn't come to pass. The event was quickly forgotten by most of those who had wondered in advance what the consequences might be.

But for those paying attention to technology trends, this was a clarifying moment: a moment in which the United States and the world recognized how dependent on computers we had become, and the ways in which the global explosion of networks and data has become—along with climate change—one of the defining features of the new millennium.

Confronted with the intelligence-gathering challenges of the post-9/11 world, NSA and the rest of the US Intelligence Community had to vault a new set of hurdles in order to gather the necessary information to identify national security threats. When it came to longtime adversaries like Russia, China, and North Korea, NSA had well-grooved approaches to gathering intelligence. Military call signs and radio frequencies were known and monitored; landline telephones to offices of defense and foreign ministries had long since been identified; even official email accounts for adversary foreign governments were often known and could be targeted using conventional intelligence collection means.

But what about international terrorists? Less was known about their organization and structure. Some operated as lone wolves or as freelance cells, inspired by larger groups like al Qaeda but not necessarily reporting directly to them. These groups certainly didn't have a central headquarters complex or published lists of landline telephone numbers and names and titles of officers and employees; there were no readily identifiable office buildings where low-level functionaries reported to work each day; there were no parking lots, metro stations, or bus routes that outlined the daily travel of terrorist organization personnel. How were existing and aspiring terrorists to be identified? How would their communications be collected, processed, and analyzed when they were swimming in the swirling morass of global traffic that had come to characterize the internet?

When I began work at NSA in 2003, the US Intelligence Community was still in the throes of self-reflection over the fact that it had been unable to predict and prevent the al Qaeda attacks on the United States that had taken place on 9/11. Congressional committees excoriated the IC. A bipartisan commission had been established to investigate the IC's failings and to make recommendations for reform. Those recommendations would eventually lead to new legislation reorganizing the IC and granting it new authorities. In the meantime, NSA Director Michael Hayden made sure his workforce knew the challenge that we faced: Our new adversaries, international terrorists, weren't communicating in the same old static ways that nation-states had communicated in the past. Instead, the chatter between individual terrorists and among their groups was riding on the same tele-communications backbone as the innocent conversations of everyone else. The terrorists were using free webmail services like Yahoo and Gmail. Their leaders were posting radicalization videos on YouTube. They used web forums in much the same way that video gamers, chess enthusiasts, and countless other shared-interest groups did.

For NSA to separate the wheat from the chaff, and identify the signal within the noise, meant that we had to deal with the problem of the volume, velocity, and variety of communications that were taking place online. And these challenges were occurring against a backdrop in which the congressio-nal "Joint Inquiry into Intelligence Community Activities before and after the Terrorist Attacks of September 11, 2001" would levy a number of criti-cisms against NSA for being overly conservative in carrying out its mission. The Joint Inquiry commission concluded that:

> Prior to September 11, the Intelligence Community's ability to pro-duce significant and timely signals intelligence on counterterrorism was limited by NSA's failure to address modern communications technology aggressively, continuing conflict between Intelligence Community agencies, NSA's cautious approach to any collection of intelligence relating to activities in the United States, and insuffi-cient collaboration between NSA and the FBI regarding the poten-tial for terrorist attacks within the United States.[1]

The report also noted with disapproval that:

> Consistent with its focus on communications abroad, NSA adopted a policy that avoided intercepting the communications between individuals in the United States and foreign countries.
>
> NSA adopted this policy even though the collection of such communications is within its mission and it would have been possi-ble for NSA to obtain FISA Court authorization for such collection. NSA Director Hayden testified to the Joint Inquiry that NSA did not

want to be perceived as targeting individuals in the United States and believed that the FBI was instead responsible for conducting such surveillance.[2]

In other words, in the years since the Church and Pike Committee hearings, NSA had taken seriously its mandate to stand at the shores of the nation and look out, to focus its intelligence collection activities overseas, and to scrupulously avoid intelligence activities that could impact Americans.

In response to these criticisms, Director Hayden testified to the Joint Inquiry:

> The volume, variety and velocity of human communications make our mission more difficult each day. A SIGINT [signal intelligence] agency has to look like its target. We have to master whatever technology the target is using. If we don't, we literally don't hear him; or if we do, we cannot turn the "beeps and squeaks" into something intelligible. We had competed successfully against a resource-poor, oligarchic, technologically inferior, and overly bureaucratic nation state. Now we had to keep pace with a global telecommunications revolution, probably the most dramatic revolution in human communications since Gutenberg's invention of movable type.[3]

The NSA's attempts to grapple with those challenges—to find adversaries' communications in the swarm of global data, and to do so without trampling on innocent individuals' rights—brought its activities smack into a set of controversies that led the organization to be accused of "dragnet," indiscriminate collection. Among the important issues that critics often seemed to overlook: First, the real and pressing question of what the alternative might be—that is, of whether there were other, less intrusive but similarly effective means to carry out the nation's intelligence work. Second, if some degree of data collection was required, and some amount of irrelevant communications would be nearly impossible to avoid, what set of post-collection mechanisms could assuage civil liberties concerns? The criticisms were seldom teased out in such precise terms. Yet they've led to extensive debate over just how much data collection should be permissible in the pursuit of foreign intelligence, and what the scope of statutory foreign intelligence surveillance authorities should be.

CHAPTER 16
MASS SURVEILLANCE AND BULK INTERCEPTION

A Distinction with a Difference

P rivacy advocates often use terms like "dragnet" and "mass surveillance" to condemn government surveillance programs but not always with much rigorous analysis behind them. In understanding the scope of US intelligence electronic surveillance programs over the past twenty years, as well as the likely future directions of border patrol and law enforcement agencies, it's useful to look at what "bulk interception" really means, and why—even within large-scale data collection programs—there are meaningful differences in the ways that mass surveillance and bulk interception can impact individual privacy. The topic is particularly timely in light of the US Supreme Court's 2018 decision in *Carpenter v. United States*,[1] which places limits on the government's ability to gather customers' location data from mobile phone companies—but which offers very little practical guidance for other metadata collection programs in the future.

THE PRESIDENT'S
SURVEILLANCE PROGRAM

In December 2005, I sat with friends and colleagues at a chain restaurant in Columbia, Maryland. I had been working at the National Security Agency for over two years and had recently moved from my original position working on information sharing and related issues in the NSA's Office of Policy to a new role as an attorney in the Operations group of the NSA Office of General Counsel. The variety and complexity of questions, and the importance of the work, had already made this the most satisfying legal job I'd held. Although the work changed constantly, we often spent our days reviewing the evidence supporting FISA applications and responding to questions from intelligence collectors and analysts about how the requirements of USSID 18—NSA's minimization procedures for activities carried out under EO 12333—should be applied to a particular operational activity. We delivered the in-person training on NSA's legal authorities and restrictions that was required annually for anyone whose job required them to access "raw" SIGINT information. I had also been assigned the new technology capability portfolio, providing legal advice on the tools NSA was developing to keep pace with the changing telecommunications environment.

On this particular afternoon, we were having our office holiday party lunch. Time away from the office was precious and hard to get: in an agency filled with military and civilian personnel whose highest priority was our shared mission of protecting the nation, we all worked extended hours, and regularly missed dinnertime at home or stayed late into the night or came in on weekends to meet whatever operational crisis was brewing at the time. After all, intelligence needed to be gathered 24/7, geopolitical crises happened in every time zone, and US and allied military forces needed battlefield intelligence everywhere they were around the world.

Unlike most other twenty-first-century jobs, at NSA we had no ability to take our work home with us. We couldn't log in through a VPN from our home computer or take stacks of paperwork home for review. By definition, almost everything we did was classified—and that was certainly true of the Ops lawyers, whose entire job was to advise the agency and its personnel on its intelligence operations. Whether we were having in-person meetings, phone calls, or emails with NSA colleagues or with overseers from the Department of Justice or Congress, or with counterparts from other IC agencies or other parts of the Department of Defense, very nearly everything we did was done via secure communications. Our meetings were held indoors in Sensitive Compartmented Information Facilities (SCIFs), our emails stayed within NSA's classified NSAnet system or traveled via a top-secret network known as JWICS, and our phone calls were carried out on the National Secure Telephone System (NSTS). We barely bothered using unclassified phone lines or email addresses, and we certainly never carried cell phones inside any of NSA's classified facilities; mobile devices were prohibited inside SCIFs, as the information security risk was far too great. And although the NSA watch centers typically had television screens tuned to a variety of news stations, when we were in our offices, we were largely cut off from outside events.

As we sat down over salads and iced tea, the news of a breaking story in the *New York Times* traveled slowly around the restaurant, like stadium seating experiencing a slow-motion wave.

According to the *Times* report,[2] President Bush had signed an order in 2002 authorizing NSA to intercept the international telephone calls and emails of "hundreds, perhaps thousands," of people inside the United States without warrants. The article contained an extraordinary level of detail in its allegations regarding the program and the views expressed by current and former government officials on where the legal line might fall with respect to these activities.

As the conversation spread over lunch, I was one of many who wondered what this might mean. Within NSA and other intelligence agencies, there was a long-standing practice of compartmentation—of holding the most sensitive information within specially designated channels such that the information was only accessible to a small group of individuals. A person had to have a Top Secret clearance to get hired at NSA; but that didn't mean each person had access to everything. If information was formally compartmented, it could only be accessed by the named individuals who had been "read in" to the compartment. Consequently, many of us at that lunch had no firsthand knowledge about how accurate the allegations in the *Times* article might be. But it didn't take any insider knowledge to realize that the article's publication was likely a watershed moment in NSA history. It felt weighty, like we were on the brink of change.

Change did, indeed, come. At first it took the form of more newspaper articles, and privacy advocates decrying the use of "warrantless wiretapping" while glossing over the fact that the Constitution doesn't in fact require a warrant for searches to be lawful.[3] By January 2007, the program had been brought within the FISA framework through orders signed by a FISC judge.[4] By April 2007, the Director of National Intelligence had submitted to Congress proposed legislation to bring FISA up to date with the changes in communications technology, preserve the privacy interests of persons in the United States, and secure assistance from private entities by granting them limited immunity for liability for cooperating with the government.[5] Congressional hearings followed soon after, and by May 2007—with lightning speed, as legislative timescales go—Congress had passed a new law, the Protect America Act, a stopgap measure valid for a mere six months that would create a statutory framework for something approximating the program outlined in the *New York Times* article. In the meantime, dozens of lawsuits had been filed seeking billions of dollars in damages from telecommunications service providers over their alleged unlawful cooperation with US government intelligence requests. A number of other lawsuits were filed directly against the government, alleging that the surveillance program violated the Constitution as well as various provisions of FISA.[6]

FISA MODERNIZATION

In the wake of the lawsuits and continued negative coverage of the Terrorist Surveillance Program, and against a backdrop of continued international terrorist activity like the 2006 airline shoe bomber in the UK and a pair of 2007 terrorist bombing plots in Germany, as well as ongoing terrorist threats to the United States, Congress felt a keen pressure to act. In order to do this, however, the government would have to be more forthcoming than it ever had before in describing why the current laws needed to be updated, and what was at stake. Admiral Mike McConnell, the Director of National

Intelligence, explained to Congress that "FISA was enacted before cell phones, before e-mail, and before the Internet was a tool used by hundreds of millions of people worldwide every day."[7]

If hindsight is 20/20, it was easy to see, in 2007, that the way the law had been drafted thirty years earlier had been a mistake. The 1978 legislation contained complicated provisions directing the government to seek court authorization for certain kinds of surveillance, depending on a combination of factors including whether the collection would be done via radio signal or from a wire, whether it was to be effectuated in the United States or overseas, and the location and suspected nationality of the communicants. Those particular provisions had one intention: to protect the Fourth Amendment rights of US persons and persons (of any nationality) in the United States, while allowing the government more latitude in carrying out electronic surveillance when it was directed at non-US persons who were overseas. This was a perfectly sensible balance of individual rights and national security, and one that is typical for most Western democracies. But in the decades since the law had been passed, changes in telecommunications infrastructure had turned the law's impact on its head. Arcane references to "radio" and "wire" communications had a different impact with a differently architected global communications infrastructure. Now, McConnell, explained, "FISA's requirement to obtain a court order, based on a showing of probable cause, slowed, and in some cases prevented altogether, the Government's ability to collect foreign intelligence information, without serving any substantial privacy or civil liberties interests."[8]

McConnell's point was that changes in technology had flipped the script in a way that meant the government's legitimate intelligence-gathering activities were significantly restricted but for reasons that did nothing to enhance privacy protections for US persons. In order to get back to the original legislative purpose, and to a public policy that was more in keeping with the Constitution's ideals, it was necessary to update the law. And all of this was required because the law was rooted in the specific technology of one point in time, over a quarter century before.

This debate marked an extraordinary moment in intelligence history, and especially in the history of electronic surveillance. The techniques used to gather signals intelligence information have always been closely guarded, because they are so perishable. The precise details were frequently among the most highly classified information at the NSA, and NSA had long been regarded as one of the most secretive agencies in the IC. If intelligence targets know what kinds of communications can be intercepted, they might change tactics and their communications would avoid detection.

Consequently, it was hard to say which was more striking: the shock felt by privacy advocates when the 2005 *New York Times* story broke, or the astonishment felt by NSA personnel when they heard their director and other IC leaders speak so openly, so publicly, about the ways in which FISA targeting worked, the shortfalls of the current legislation, and why the current threat environment required changes in the law.

That's how important this was inside the Intelligence Community; that was the severity of the threat. Changing the law, modernizing FISA, wasn't a pedantic exercise carried out to achieve some greater level of bureaucratic need—it was a matter of life and death for the safety of the nation. Nothing less than that could have justified the decision to reveal so much, at a time when everything in the culture of the IC was steeped in a long tradition of reticence and caution for fear that tenuous intelligence collection capabilities might completely perish.

Ultimately, Congress agreed with the administration's rationale, replacing the Protect America Act with the FISA Amendments Act, a more enduring piece of legislation with a six-year sunset built into the law. The FAA has since been extended three more times, with reauthorizations in 2012, 2017, and 2019.

During the 2017 reauthorization cycle, I had the privilege of testifying before Congress on the nature and effects of the law, in particular section 702, which allowed the IC to target the communications of a non-US person outside the United States for foreign intelligence reasons without obtaining a probable-cause order from the FISC. Section 702 and the court-ordered targeting and minimization procedures associated with it were carried out under extensive scrutiny and oversight. And the program had amassed a considerable track record across all of the dimensions that mattered most: intelligence value, impact to privacy, and the way that overseers handled missteps by the

IC when information was handled incorrectly. In 2014, an independent and bipartisan agency, the President's Civil Liberties and Oversight Board (PCLOB), had conducted an exhaustive review of FISA operations within the IC. In my congressional testimony, I wrote:

> The first and most important point to make is that, despite some public misconceptions to the contrary, FAA 702 is a targeted intelligence authority. It is not "bulk" collection. As explained by the independent Privacy and Civil Liberties Oversight Board (PCLOB) in its July, 2014 report, "The statutory scope of Section 702 can be defined as follows: Section 702 of FISA permits the Attorney General and the Director of National Intelligence to jointly authorize the 1) targeting of persons who are not United States persons, 2) who are reasonably believed to be located outside the United States, 3) with the compelled assistance of an electronic communication service provider, 4) in order to acquire foreign intelligence information."[9]
>
> In more concrete terms, FAA 702 collection can only be initiated when an analyst is able to articulate, and document, a specific set of facts to meet the statutory and procedural requirements for demonstrating that: 1) a specific "facility" (such as a phone number or email address) 2) is associated with a specific user 3) who is a non-US person 4) who is reasonably believed to be located outside the US and 5) who is likely to possess or communicate foreign intelligence information.
>
> Although a large number of selectors have been targeted under FAA 702, each of those selectors has been tasked for collection because on an individual, particularized basis each one of them meets the criteria noted above. "Bulk" collection is different: as explained in "Presidential Policy Directive – Signals Intelligence Activities" (PPD-28), bulk collection is information that is collected without the use of discriminants. This is a critically important difference . . .
>
> Further, once the information has been collected under FAA 702, the information is subject to a significant number of post-collection

safeguards that are captured in lengthy, detailed minimization procedures that demonstrate both the care that is taken with the information, and the complexity of the 702 framework.[10]

Many critics of 702 have raised concerns about incidental collection: the inevitable reality that there are times when foreign intelligence targets communicate with US persons, and the government's collection of targeted communications incidentally pulls in the part of the conversation that includes the US person.[11] In thinking through the privacy impact of incidental collection, it's worth remembering that, from the law's beginning in 1978, Congress was aware of this risk. In the original legislative history, Congress viewed this as an appropriate and acceptable trade-off. As far back as the 1970s, federal courts had approved the Constitutionality of incidental collection.[12] Congress specifically noted, in its discussion of how minimization procedures should be designed, that there was no obligation for the intelligence agencies to destroy incidentally collected information that might have foreign intelligence value simply because it might also involve a US person.[13] Rather than imposing an outright prohibition on collecting or retaining that information, Congress favored a more balanced approach, the principles of which have remained intact to this day. The statute and court-ordered procedures "minimize the acquisition, retention, and dissemination of information" relating to US persons, "consistent with the need of the United States to obtain, produce, and disseminate foreign intelligence information."

Perhaps as an expression of that concern, or perhaps simply to understand the scope of the challenge, a number of voices in the privacy advocacy community, as well as in Congress, have called for NSA to count, document, and publish the number of US person communications that are swept up in its FAA 702 collection. NSA has consistently taken the position that this would be both impossible and unwise. On this point, NSA is correct.

Although I spent many years at NSA, I was fully independent of it when I testified as a witness before the House Judiciary Committee during hearings it held on the reauthorization of FAA 702. In that testimony, I tried to convey why it would be unwise to impose on the Intelligence Community an obligation to identify all communications to, from, or about US persons:

From a policy and privacy perspective, the current approach—in which analysts only research unknown identifiers when they appear likely to be of intelligence interest—is a sound and sensible one that protects privacy, conserves resources, and helps the government focus on the highest intelligence priorities. A requirement to count the number of US person communications that are incidentally acquired under Section 702 would require the Intelligence Community to conduct exhaustive analysis of every unknown identifier in order to determine whether they are being used inside or outside the US, and whether their users might be US persons located anywhere in the world. NSA does not—nor should it—collect or maintain comprehensive directories of the communications identifiers used by US persons. However, in order to perform a reliable count of US person communications in 702 collection, the Intelligence Community would have to create and maintain precisely such a database. The very creation of these reference databases would constitute an unnecessary and unwarranted intrusion on the privacy of US persons; without specific statutory authorization, it would likely also be unlawful, since it would be both intrusive and unrelated to any need for foreign intelligence gathering. (Even with statutory authorization, the creation of such a comprehensive database would raise Constitutional concerns.)

Further, searching for US person information would require intelligence agencies to divert scarce analyst time and computing resources away from intelligence activities in order to hunt for the communications of US persons whose information is not related to an authorized intelligence need (and whose information would never be looked at by the government but for this requirement). Finally, it is unlikely that knowing the number or percentage of US persons in a particular sample of data would result in increased privacy protections in the future: first, because target sets vary over time, and therefore it isn't clear whether numbers or percentages of incidental collection would be constant over time; and second, because the fundamental challenge remains an intractable one: as long as foreign intelligence targets communicate with US persons, it will not be possible to avoid the incidental collection of those specific communications. The best way to protect the privacy of incidental US person communications is to advise analysts that they should not proactively search for communications that lack intelligence value, nor conduct exhaustive research to determine whether the unknown communicants in irrelevant communications might be US persons or persons in the US.[14]

If national security is one of the first and most basic jobs of the federal government, then the United States will continue to need access to intelligence information, and to the investigative tools that make it possible to detect, prevent, and respond to threats, whether those threats arise from foreign actors or from within the country's borders. When the nation's three branches of government are functioning in robust and effective ways; when the press acts as an effective fourth estate, examining and reporting on national security activities; when informed citizens express their views, directly or through advocacy groups, history has shown that our legal frameworks, norms, and traditions provide the tools we need to keep government accountable, to set proper boundaries around intelligence activities, and to hold officials accountable when they overreach. The challenge for the future is bearing in mind the lessons of the past, and making sure that none of these crucial mechanisms break down.

CHAPTER 17
COMMUNITY POLICING

All Surveillance Is Local

M uch of the attention to government surveillance in the past decade has focused on national agencies like those in the US Intelligence Community. However, a great deal of government surveillance takes place at the state and local level, where regulations, and the standards for enforcing and overseeing them, can vary widely from one jurisdiction to the next.

POLICE USE OF
SURVEILLANCE TECHNOLOGY

Police around the nation have, for years, intercepted radio signals as a way to identify mobile phones in the area. Generally, they use a category of tools known as Stingrays, or cell-site simulators.[1] Stingrays work by masquerading as a cell phone tower, causing any nearby cell phones to respond to the Stingray's signal ping. In some cases, police use these for general investigative purposes. In other cases, they are searching for a specific phone and using the Stingray to suss out whether its owner is nearby.

Traditionally, the Fourth Amendment hasn't regulated the collection of radio signals for the simple reason that radio transmissions are, by their nature, publicly available: anyone with an antenna and a receiver can intercept a radio signal as it flies by. The equipment is inexpensive, and the science is simple. As a result, radio signals have not been viewed as carrying the same expectation of privacy as communications carried across, say, a fiber-optic cable or a telephone wire. Stingray pings were thought to require even less protection, as there is no content being conveyed, only the fact that a phone is in a location. Despite this low expectation of privacy, a number of courts around the nation have held that a warrant is required before police can use a Stingray device; the Supreme Court, however, has not yet taken up a cell-site simulator case. Police will almost certainly continue to use them, but under legal standards, and levels of restriction, that vary from one state to another nationwide.

In 2018, the city of Newark, New Jersey, invited local residents to watch surveillance video from public street cameras and help the police in solving crimes. Previously, the video feeds had been accessible only within the police department's closed-circuit TV system. Now, they would be accessible to anyone with an internet connection. Police viewed this as having concrete as well as intangible benefits: local residents might be able to identify individuals who were unfamiliar to police, and the citizen-officer cooperation might foster a sense of partnership in a city with a history of fraught relationships between residents and the police. Privacy advocates worried about similarly concrete and general concerns: would-be stalkers or burglars might use the feeds to monitor the behavior of people who lived or worked near a particular set of cameras, and residents who reported potential identities to the police would not have had the benefit of department-sponsored training on issues like racial, ethnic, and religious discrimination and the role of unconscious bias.[2]

Police use of body cameras has attracted similar kinds of opposition and support. Some opponents fear the footage they capture could intrude on the rights of people who happen to be passersby at the time of a police action; other opponents also argue that the cameras are unnecessarily intrusive on the rights of the officers themselves, forcing them to work under constant scrutiny. Supporters of body cams make the opposite civil liberties points: in jurisdictions where body cameras are required, there's been an increase in the number of internal investigations and prosecutions of police officers who were caught on their own cameras planting drugs, beating up suspects, or taking other actions that would likely not have been caught had it not been for the evidence from the body cams.[3]

Police nationwide are also turning to drones to monitor crowds. After the 2017 mass shooting in Las Vegas, the local police department announced that it would deploy drones to monitor crowds during that year's New Year's Eve celebration. The camera-equipped drones could help identify suspicious packages, track unusual behavior, and perhaps help police make sensible adjustments to street barriers and pedestrian flow along the Strip if needed. The police had, ironically, purchased the drones just before the October shooting, but the devices weren't yet ready to fly. If they had been, they might have helped identify victims in need of assistance, and even create a real-time 3D model of the crime scene, perhaps even helping the police more quickly identify the shooter's location.[4]

Despite the potential benefits for law enforcement, police use of drones creates a number of risks. Critics and supporters agree that drones provide extraordinary access: they can navigate in narrow, congested locations inaccessible to manned aircraft, and their optics can routinely deliver high-resolution imagery with a hyper-powerful 180× zoom. On top of that, they're relatively cheap: a fully kitted-out drone can be bought for as little as $6,000, compared with police helicopters whose cost can easily reach $3 million.[5] In addition to the risk of overly intrusive data collection, there are also very real concerns that drones could be used in ways—like flying over protests or demonstrations, or outside houses of worship—that could dissuade people from exercising their First Amendment rights to freedom of association, speech, and religion.[6] This combination of low cost and highly effective surveillance is likely to push more police departments toward using drones, often before police procedural guidelines have been updated to incorporate appropriate protections for privacy and civil liberties.

COMPUTERIZED DATA AND INDIVIDUAL RIGHTS

Congress and the executive branch have long recognized the power and dangers of data aggregation and automated analysis. It's why Congress, long beholden to scorecards, fundraising, and lobbying efforts by the National Rifle Association, has passed laws prohibiting the creation of any computerized databases of gun registration records, and prohibiting the Centers for Disease Control from investigating gun violence as a public health concern. Although legislators' restrictions on gun registration databases are couched in terms of defending Second Amendment rights, the restrictions have been adopted and enforced precisely because of the recognition that, when the government holds vast quantities of computerized information about its citizens and residents, it has in its possession a tool that starts with privacy-intrusive information that can quickly be martialed to infringe on other rights.

For right-leaning politicians, the privacy intrusion threatens Second Amendment rights. For left-leaning politicians, privacy intrusions threaten First Amendment rights to protest government action and to advocate for civil rights and advancement of civil liberties. Not only is information about individual gun registrations not computerized, but other types of data about gun usage—statistical information on injuries, fatalities, demographics of gun owners, trends in gun-related accidents, etc.—are seldom used in large-scale data analysis. Some of this is due to long-standing rules limiting the use of federal funds to support gun-related public health research at agencies like the Centers for Disease Control. It also results from statutory constraints lobbied for by the NRA. As a result, the National Tracing Center, which receives requests from law enforcement agencies all over the country to trace firearms that were used in a crime, has to deal with everything manually. On paper. And sometimes on microfilm. One million new gun registrations a month. Fifteen hundred active gun traces on any given day. Three hundred and seventy thousand gun traces per year. All of it done without computers. This slow-motion manual process creates frustration for police, for the Alcohol, Tobacco, and Firearms officers who are assigned to the Tracing Center, and probably causes frustration for many gun control advocates as well, who would like to see research carried out on trends relating to gun ownership, types of gun used in relation to types of crimes, and other sorts of statistical analysis that might bolster the opportunity to suggest evidence-based gun laws.[7]

Regardless of their stance on the availability of guns and the ways in which the Second Amendment should be harmonized with modern American life, almost anyone who looks at this arrangement is bound to shake their heads and say, "Yep, if you want to make it hard for the government to identify who has guns or interfere with their ability to own them, having a massive amount of information stored in cardboards boxes for manual review has got to be a good way to do it." And although the Supreme Court's recent decisions on big data have been more focused on the rights protected under the First and Fourth Amendments, in many respects the upshot is the same: computerization of information makes it far too easy for government and corporations to learn too much too quickly about each one of us for anybody's comfort.

Finally, in a striking intersection between surveillance by private individuals, private-sector platforms, and police, some four hundred police departments nationwide have contracts with Amazon that allow the police, without probable cause or a warrant, to review the videos that are uploaded every day from the thousands of individuals who've installed Amazon's video-surveillance doorbell, called Ring.[8] The Ring doorbell was created by Amazon as a way for people to record video footage of anyone who approaches their door—a commercially beneficial product for Amazon, whose retail business depends heavily on the willingness of its customers to have packages delivered to their homes. Ring footage is stored on Amazon cloud hosting. When customers install Ring, they also download an app called Neighborhood, which lets them view that footage and which has a social media component allowing Ring owners to post messages to other residents in their area. This feature is advertised as providing something similar to a virtual neighborhood watch program. Ring users whose packages have been stolen from the front steps, or who have video showing other crimes, can report these to the police, and the footage may prove to be a useful investigative tool.

Under these contracts, however, police don't have to wait for a resident to report a crime; police can proactively search the Ring footage archives for information relating to a criminal investigation. Since they don't need a warrant to review this information, it's not hard to imagine that the access could be misused. One Massachusetts lawmaker has blasted Amazon's product over privacy concerns, stating, "Connected doorbells are well on their way to becoming a mainstay of American households, and the lack of privacy and civil rights protections is nothing short of chilling." In a series of tweets, Senator Ed Markey wrote that Ring's policies don't impose security requirements on police that access the videos, don't restrict police from further sharing the videos with third parties, and don't include policies that would limit how long police can hold on to the video clips.[9] The potential privacy risks have been complicated by viral videos and news stories about the ease with which unauthorized users can hack into Ring devices, controlling the video feed and even carrying on chilling conversations with children whose parents installed Ring as a home safety measure.[10] Police could make a habit of viewing videos in neighborhoods with high percentages of minority residents, for example. They could persistently search video of a particular individual who they want to target, whether for legitimate reasons or out of a vindictive desire to catch them at something that could be charged as a crime. They could also stumble across evidence of a crime that might otherwise have gone uncharged, such as when police serendipitously view footage of what appears to be drug use taking place within sight of a Ring camera. Ironically, the footage could conceivably be used against the Ring owner themselves, most likely not an outcome that anyone hopes for when installing these home surveillance devices.

PRIVATE MISUSE OF SURVEILLANCE TECHNOLOGY

On a July day in 2019, the most-viewed video on the BBC News website was a short piece about a California neighborhood that had installed cameras on light poles to capture the license plates of vehicles entering and leaving the community. Community residents had been concerned about a rise in local crime—so they turned to a company that provided a system of pole-based cameras and cloud storage of the imagery showing traffic through the neighborhood at modest prices: $10 or $20 per month added on to their community association dues. The program had a number of built-in safeguards: still images were only stored for thirty days, all data was password-protected, only four designated users from the community association were authorized to access the data, and the company even had its customers sign a contract indicating they understood that the information collected and stored by its system was to be used solely for purposes of criminal investigation, not for surveillance.

Despite these safeguards, a privacy professor interviewed for the BBC News piece suggested that perhaps it could be misused: a jealous partner might use it for stalking, or it could open the door to other, more intrusive technologies in the future. I could see the scenarios for misuse in my own community, where a handful of neighbors zealously sought to enforce the posted speed limit at all times. It wasn't hard for me to imagine a neighbor shaking their fist at the white pickup truck that sped through the neighborhood precisely at 3:15 PM on a Tuesday afternoon. They could then demand that the community association look at the still images to identify the offender's license plate and take some suitably punitive action.[11]

Although an overzealous or vindictive neighbor *could* misuse this technology, such scenarios seemed unlikely when compared with the obvious, intended use: recording road traffic so that the community association could give the police vehicle information that might be relevant in investigating reported crimes. To be sure, personal data can be alluring to private citizens, who are just as prone as governments and corporations to exercising the worst tendencies of human nature. At the same time, we shouldn't shy away from innovative programs that might improve quality of life. Instead, we should look for examples where those activities seem well designed and sensibly bounded, and incorporate privacy-protective measures in ways that satisfy a community's goals.

THE FOURTH AMENDMENT AND SURVEILLANCE TECHNOLOGY

There are several strains of Fourth Amendment law that support what has long been a generally held, commonsense view about governmental intrusion: if something happens in an open field or in plain view, then police don't need a warrant to investigate what they can see.[12] The idea is that things that are obvious and that happen in the open aren't entitled to special protection under the Fourth Amendment. There's another legal theory known as the third-party doctrine, which says that if a person has already entrusted their information to someone else, then the police don't need a warrant to obtain that information directly from the third party. Here, the rationale is that, because it's already been shared, the information is no longer private, and if a person discovers that they've placed their confidence in someone untrustworthy, that is no fault of the police and requires no remedy from the courts.[13] Under all three of these conditions—open field, plain view, and the third-party doctrine—courts have repeatedly held that it's reasonable for police to search or seize people or things (including information) without meeting the "probable cause" standard of proof. As a result, the Fourth Amendment doesn't require police to obtain warrants in these circumstances.

Data-driven technologies and the wealth of information they can accumulate have strained these legal theories. In recent years, the Supreme Court has decided a handful of cases addressing the circumstances under which modern technology does, or doesn't, modify the open-field and plain-view doctrines. During the 1980s, the court held that aerial photography didn't require a warrant, even when police were flying over a person's home or business searching for incriminating information.[14] Under this line of cases, if evidence of criminal activity was visible to the naked eye, then it didn't matter, from the standpoint of Fourth Amendment protections, what angle the police viewed the evidence from, and whether that angle of approach was assisted by helicopters or fixed-wing aircraft. As technology became more sophisticated, however, the court's opinions showed that their rationales were less clearly defined; in some cases, it was simply apparent that the court found the facts troubling and felt that some remedy was required, even if they couldn't articulate a coherent theory as to why.

One of the first cases that showed the court struggling to reconcile the Fourth Amendment with new technology involved police using thermal imaging to detect heat signatures emanating from a man's home.[15] In 1991, federal agents suspected Danny Kyllo might be growing marijuana indoors. His electric bills were unusually high (which could be caused by the high-intensity grow-lights required for indoor cultivation). They parked a car across the street from his house, pointed a thermal scanner at his home, and determined that the heat signatures were, indeed, anomalous. Armed with this information along with previous investigative work, they sought a warrant to search his property, and discovered that he was growing dozens of the illicit plants. Kyllo was convicted of marijuana cultivation and appealed, arguing that the thermal scan had violated his Fourth Amendment rights, and that therefore all subsequent evidence was tainted and should have been dismissed by the lower courts.

Nearly a decade later, when Kyllo's case reached the Supreme Court, the justices agreed with him. At the heart of their decision was the idea that, "Where, as here, the Government uses a device that is not in general public use, to explore details of the home that would previously have been unknown and unknowable without physical intrusion," Fourth Amendment protections apply. In other words, the police should have gotten a warrant before turning the thermal scanner on Kyllo's home.[16] What's striking here is that the thermal scan was carried out in 1992; by the time the court issued its decision in 2001, thermal imagery had become far more readily available. And nearly twenty years later, thermal imaging techniques are inexpensive and common. The *Kyllo* decision raises the question: What is the shelf life of that window when a new technology is sufficiently rare that it meets the "not in general public use" factor of the court's test? In an era where Moore's law predicts that data-driven technologies are likely to continue becoming twice as advanced every eighteen months to two years, the availability, cost, and commonness of use for any particular technology could change half a dozen times before a case makes it through the courts.

The difficulty with making technology-based decisions remained on display a decade later, when the Supreme Court decided *United States v. Jones*.[17] The case began in 2005, when a joint FBI and Metropolitan Police Department task force suspected Antoine Jones of engaging in drug trafficking. They deployed a number of surveillance techniques; among them, they attached a GPS device to his car and tracked his movements for the next twenty-eight days. It's helpful to put this case in the context of its time: GPS devices weren't routinely incorporated into vehicle telematics; after-market GPS for cars was still a relatively boutique item; smartphones didn't exist; and, as a result, we hadn't yet become culturally accustomed to a world in which our whereabouts are being tracked by devices on a 24/7 basis.

Jones argued that this continuous tracking of his whereabouts should have required a warrant. The Supreme Court agreed. The court's decision in this case is a complicated one, with multiple opinions and different members of the court concurring in the outcome for different reasons. At the heart of their disagreement was whether the Fourth Amendment violation was based on property ownership and trespass principles from eighteenth-century common law, when the Fourth Amendment was written,[18] or the twentieth-century "reasonable expectation of privacy" test that the court had articulated for electronic surveillance.[19] Five of the justices believed that changing technology required different approaches than the early history of constitutional law.

In one of the concurring opinions, Justice Samuel Alito noted the uncertainty that can occur when new technology prods uncertain shifts in cultural norms. Further, he said, "Society's expectation has been that law enforcement agents and others would not—and indeed, in the main, simply could not—secretly monitor and catalogue every single movement of any individual's car for a very long period of time." The *Jones* case showed the fractures in Fourth Amendment legal theory, and the more recent *Carpenter v. United States*, in which police gained access to cell-site location information without a warrant, only resulted in a further tangling of these different theories of law. Although some cases, like *Jones* and *Carpenter*, reach a single result based on multiple reasons, at some point the court will have to develop a coherent approach that doesn't stay mired in the fiction that eighteenth-century life translates perfectly to today; that doesn't cause the legal standard to change as soon as a particular kind of technology changes status from emerging to commonplace; and that wrestles with the modern digital reality that nearly all of our data is hosted by, or visible to, someone else.

LOCAL RULE AND LOCAL RULES

In communities around the country, some residents have expressed a desire to see modern technology serve as a force multiplier for local police in their efforts to fight crime. Others have articulated nagging unease about the ways in which surveillance technology might be used to create a society that is inhospitable to core privacy values, such as personal autonomy, freedom to behave anonymously, the ability to make limited disclosures of information, and the like. In response, San Francisco became the first city in the United States to ban the use of facial recognition systems in street cameras, citing the racial bias and inaccuracy of many such systems, and the corresponding disproportionate effect they can have on communities of color.[20] Since that decision in May 2019, a number of other cities have followed suit.

Unlike the federal government, most state and local jurisdictions have little if any legislation governing the collection and use of this information. As a result, each police department is largely left to its own devices to set policies regarding what kinds of technologies to adopt, what kinds of information to collect and in what circumstances, who may access the data and why, how long the information may be retained, and how it may be used. Harkening back to the lessons on oversight from the Church and Pike Committees, public safety remains an undeniably vital service that government has a duty to its citizens to provide, and in many instances, the danger of overreach stems at least as much from overzealous and misguided human decisions as from the technology used to carry those decisions out. Those risks can be managed through clear laws, procedures, oversight, and guidance; but leaving these activities unregulated amplifies the risks that surveillance technologies will be misused. Instead of insisting that new technologies not be used, community members and privacy advocates alike should press for clear regulatory schemes that are made transparent to the public, a requirement for robust internal compliance programs, and empowered and independent external oversight mechanisms to make sure the technologies are being used only in ways that the citizenry has approved and that have been appropriately disclosed.

CHAPTER 18
GOVERNMENT SURVEILLANCE IN A TIME OF TRUMP

Why We Still Need It, How to Control It,
and How to Protect Ourselves Against It

T he Trump administration has stretched the bounds of executive power in ways that are deeply disturbing to privacy advocates and promoters of civil liberties throughout the country and around the world. Thus far, Congress hasn't taken any significant actions to rein him in. The question is: Regardless of how long Trump remains president, and regardless of who controls Congress and the White House in 2020 and beyond, what have recent years shown us about the risk of governmental overreach, and the appropriate balance between privacy and surveillance?

The record is mixed.

Fearmongering aside, the United States isn't an authoritarian surveillance state—that distinction goes to countries like China and Russia, whose surveillance programs are described in later chapters of this book. Those who warn that the US government could become totalitarian are right to always be on guard against the erosion of individual liberties. However, the mere possibility that the balance between state power and individual autonomy could shift in ways we're uncomfortable with shouldn't, by itself, lead any of us to panic over whether the deep state is surveilling conservative political campaigns (it's not) or whether there are any constraints at all that could prevent improper government surveillance of US persons (there are). Although these and other misdeeds are *possible*, and although the rapid expansion of technology and digital footprints makes abuses more possible than ever, we still have a set of national political tools to prevent, detect, correct, and punish and wrongdoing by American officials. We need to remain alert and vigilant. We need to demand government transparency and hold government leaders to account. But with an informed citizenry and sufficient political will, we can avoid becoming a surveillance state.

WHAT YOUR PHONE COMPANY KNOWS ABOUT YOU CAN HURT YOU: GOVERNMENT ACCESS TO PRIVATELY HELD DATA

The heart of protections against government surveillance lies in the Fourth Amendment, which guarantees the right of the people to be secure in their persons, homes, papers, and effects from unreasonable searches and seizures by the government and which requires that any warrants be specific about what is being sought, and be based on probable cause. One of the areas most steeped in Fourth Amendment law is electronic surveillance. Although a detailed discussion of that legal background and history is beyond the scope of this book, it's worth noting the ways that the Supreme Court has focused in recent years on the potentially intrusive nature of mobile phone data.

In 1979, in the case *Smith v. Maryland*, the US Supreme Court held that the government didn't need a warrant to ask the phone company for a copy of all the calls being made to or from a number associated with a stalking case. That case established the third-party doctrine discussed in the previous chapter. The result of that doctrine is that the records of what numbers we call, and when, are available not just to the private companies that collect, share, sell, combine, and use that information, but also to any government agency that approaches those companies with a legitimate request for access to, or copies of, the information captured in those private-sector records. Although the call detail records themselves contain few specifics, they can, as described in previous chapters, be aggregated with other data to create incredibly rich and detailed portraits of our personalities, activities, interests, and inclinations, along with the mere fact of who we've communicated with.

Every day, governments around the globe send requests or demands to private companies for information about their customers, subscribers, users, and vendors. This isn't inherently nefarious: governmental subpoena power has been established in law for centuries, and it serves valid and important societal purposes. Nonetheless, the changes in digital technology in the past fifty years have increased the scope and impact of information that's available in response to these government demands. In the 1970s, when *Smith v. Maryland* made clear that the third-party doctrine applied to telephone records, those records contained less information than they do now, and there were fewer opportunities for that slimmed-down data to be central in creating a comprehensive behavioral profile. Phone companies kept a record for billing purposes of the landline numbers that calls were made to and from, the duration of the call, and what area code they were in. A police officer with a phone directory could easily cross-reference those numbers to attach names of individuals or businesses to them, but those phone records only revealed calls made to or from a particular physical address. By 2016, advances in geolocation technology and the widespread adoption of mobile phones had drastically changed the scope of information that could be gathered, and inferred, from cell records. In 2012 and 2014, the US Supreme Court decided two cases that weren't directly related to the third-party doctrine, but that hinted at the court's growing unease with the intrusive power of location tracking technologies and the personal information stored on mobile phones. In 2012, the court held that police needed to obtain a warrant if, in order to track a suspect's location over time, they used enhanced technology to gather data in quantities and at a level of detail beyond what ordinary law enforcement staffing could make possible.[1] In 2014, the court held that the contents of cell phones were so sensitive and so comprehensive that a police search would almost always require a warrant.[2] Taken together, these cases presaged the court's likely future direction when it came to issues of data that could be gathered from mobile phones.

The court's concern was well placed. The frequency with which cell phones ping cell towers and generate location information makes it possible to carry out nearly continuous tracking of an individual. As the court noted in the 2018 *Carpenter v. United States* case, cell phones are with us on a near-constant basis—apparently, 14 percent of us even shower with our phones. The sheer volume of insight about us that can be derived from this 24/7 record of our location means that there must be Fourth Amendment protections for those compilations of information about where we've been. The court stated that it wasn't overturning the third-party doctrine, but also held that if police wanted to obtain more than a week's worth of cell phone location information from mobile phone company records, they needed to obtain a warrant and show probable cause to believe that the person was involved in criminal activity—a far higher burden of proof than was required for an ordinary business records subpoena. The *Carpenter* decision, which included multiple concurrences and dissents, was unsatisfying for Fourth Amendment scholars, because the court seemed to be trying to split a baby: while it didn't overturn the third-party doctrine, it also didn't offer a cohesive vision for how that doctrine might be applied in the future. And while it said that police needed a warrant in this case, it didn't clearly articulate a broader rule that could guide future police actions. The decision left open more questions than it answered, but it did make clear the court's continued unease with just how much data about each of us is out there, floating in the digital ether, and how much that information reveals about us.

THE LIMITS OF DATA ANONYMIZATION

One more note on cell phone location data. It turns out that just as the old crisscross or reverse-lookup phone directories that used to be exotic are now ubiquitous and easy to use, it's also true that phone location data can be understood in both "directions." We've seen that when our identity is attached to location information, the locations can provide rich insight into our activities. But the reverse is also true: even when location information has been stripped of any explicit identifiers, the mere collection of locations is enough to reveal precisely who we are.

To put it another way, even when phone data has been anonymized for purposes of research, marketing, or other reasons, it's still remarkably easy to figure out who we are by where we go. In 2013, a team of security researchers published a study concluding that "mobility traces are highly unique." Even if a person's name, address, phone number, and other obvious identifiers have been stripped from the data set, if the mobile phone data provides location information once per hour—and most phones capture location data far more often than that—researchers only need four data points to re-identify 95 percent of the individuals in the data set.[3]

This ability to reverse-engineer identity from location has both practical and legal implications. A number of laws, such as the European data privacy regulation discussed in later chapters of this book, impose fewer restrictions on the handling of "anonymized" personal data. The research, however, calls into question just how meaningful anonymization is. If our phones reveal intimate details about who we are, and our mere location reveals our identity, then the fact that we rely on these devices so heavily has put us in a position where we have very little privacy left.

THE LINE WHERE RIGHTS EVAPORATE? DATA COLLECTION AT THE BORDER

During the 2016 election season, candidate Donald Trump made border security a centerpiece of his campaign. From promises of a border wall to "a total and complete shutdown of Muslims entering the United States," Trump made clear that border security initiatives would form core components of his presidency. In the three-plus years since then, the White House and executive branch agencies have sought to implement that vision. In January 2017, just days after his inauguration, President Trump signed an executive order that denied visas to citizens of seven majority-Muslim countries, a ban that has been challenged in court, struck down, and reshaped and expanded multiple times in the three-plus years since then.[4] Work has variously been promised, attempts made at redirecting funding, and some work started on the promised border wall.[5]

Meanwhile, the federal agencies responsible for border control have stepped up searches of travelers arriving at the US border. Within a year of Trump's inauguration, the Department of Homeland Security released a report showing that it had searched an estimated 30,000 cell phones, computers, and other devices of people entering and leaving the United States—almost 60 percent more than the number of searches carried out the previous year.[6] Almost without exception, such border searches are carried out without a warrant. The Customs and Border Patrol officer at a given port of entry simply detains the traveler, and the traveler has little recourse. It isn't clear yet which direction border privacy jurisprudence is heading, so some background is in order.

As described in previous chapters, the Fourth Amendment prohibits the government from carrying out "unreasonable" searches and seizures. In most cases, the way that the government demonstrates that a search is reasonable is through obtaining a warrant, based on probable cause, that specifies the persons or things to be searched or seized. In the 2014 case *Riley v. California*, the Supreme Court specifically held that police needed a warrant in order to search the contents of a cell phone. The Riley case consolidated appeals from two criminal defendants, both of whom had been stopped for one infraction—in Riley's case, driving with expired registration tags—and then charged and convicted of far more serious offenses when officers making the stop found cell phone evidence that led to the convictions on more serious crimes. In both cases, the police contended that the cell phone searches fell within a long-standing exception to the warrant requirement, which allowed police to make a search incident to arrest. The court disagreed. The traditional search incident to arrest was intended to identify weapons or other safety threats to officers, and to preserve evidence that the arrestee might otherwise be able to destroy prior to its detection. According to Justice John G. Roberts, who wrote the majority opinion for the court, neither of those rationales could be applied to the search of a cell phone. Roberts wrote:

> These cases require us to decide how the search incident to arrest doctrine applies to modern cell phones, which are now such a pervasive and insistent part of daily life that the proverbial visitor from Mars might conclude they were an important feature of human anatomy.[7]

In reaching its decision, the court relied heavily on its assessment that modern cell phones "are in fact mini-computers" that "could just as easily be called cameras, video players, rolodexes, calendars, tape recorders, libraries, diaries, albums, televisions, maps, or newspapers."[8] Furthermore, the enormous storage capacity of modern cell phones increased the privacy implications associated with them. Cell phones store a wide variety of information —addresses, notes, prescription, bank statements, videos—that "reveal much more in combination than any isolated record." And with that storage, "the sum of an individual's private life can be reconstructed," including information that predates the owner's purchase of a particular device.[9] The court raised concerns about the privacy of internet searches, the information stored in mobile phone apps, and the fact that phones might contain information that is even broader than what could be found by ransacking a person's home, including information that exists literally nowhere else.[10]

"Modern cell phones are not just another technological convenience," Roberts wrote. "With all they contain and all they may reveal, they hold for many Americans the privacies of life."[11] Consequently, the court determined that the usual search-incident-to-arrest exception did not apply, and police were required to obtain a warrant before conducting a search of a person's cell phone.

With *Riley* as precedent, then, how could DHS search over 30,000 cell phones and computers at the US border in a single year? The answer lies in another line of Fourth Amendment jurisprudence and the border search exception. Since the 1880s, the Supreme Court has long held that the government's sovereign power is greatest at the border, and that in the balance of equities between national security and individual liberty, the balance tips toward the government at the border.[12] Under this broad authority, customs and border officials have nearly unlimited discretion to search packages and correspondence or to direct travelers to secondary screening at airports and border crossings.

The Trump administration, however, seemed to be taking the broad boarder search authority to new extremes. As part of that secondary screening, federal agents were now asking an increasing number of travelers to turn over their cell phones or laptops for inspection. In testimony before the House Committee on Homeland Security, DHS Secretary John Kelly confirmed that in some cases this screening would include a demand that the traveler provide federal agents with the password to their social media accounts, so that border officials could carry out an on-the-spot inspection of the traveler's social media feed.[13] DHS was also reserving the right to question travelers about their contacts, and even their ideology, prior to making a decision on whether to allow them into the country.

What many Americans may not have realized was that these intrusive measures were permitted inside the country as well, so long as they were being carried out in connection with border security. Under federal guidelines, the border search exception applied anywhere within one hundred miles of the nation's border. This expansive zone includes places like El Paso, Texas, that are often thought of as some of the main stages where border policies play out. But it also includes the entire states of Michigan and Florida, large chunks of Pennsylvania and Ohio, New England, New York City, Philadelphia, and Washington, DC, to name just a few. In fact, 65 percent of the US population lives somewhere inside the one-hundred-mile border zone.[14] While that border zone doesn't permit customs, border, or immigration officials to stop anyone and demand to see their laptop or cell phone, it does provide broad authority for immigration agents to stop vehicles—including Greyhound buses, Amtrak trains, and private cars—on suspicion that their passengers are undocumented immigrants. It also allows border patrol to set up checkpoints where they can demand proof of citizenship from the occupants of every vehicle, and send them to secondary screening facilities, much like that carried out in airports, based on nothing more than reasonable suspicion.[15]

Public outcry from civil libertarians and immigrant rights activists reached a head with a lawsuit filed by the American Civil Liberties Union (ACLU) and Electronic Frontier Foundation (EFF), challenging the constitutionality of these procedures. The suit, known as *Alasaad v. McAleenan*, was filed in the US District Court for the District of Massachusetts on behalf of eleven people—ten American citizens and one lawful permanent resident—who had been detained and had their digital devices searched while trying to enter the United States. In some instances, border officials carried out a "basic" search of the phone's contents, scrolling through photographs and other contents of the phone. The officers also carried out "advanced" searches, connecting some travelers' phones and devices to electronic search systems and, in one case, making an image of the traveler's entire laptop, along with all of his SD cards. With these advanced searches, digital information was extracted and retained in government databases, and used for comparison with new information the next time that the same individuals sought to return from international travel.

In reviewing *Alasaad v. McAleenan*, the US district court acknowledged the long-standing principle that government authority was at its zenith at the border. However, the court noted, "Even under the border search exception, it is the privacy interests implicated by unfettered access to such a trove of personal information that must be balanced against the promotion of paramount governmental interests at the border."[16] The district court pointed to the Supreme Court's decision in *Riley* as articulating the "vast privacy interests" against which governmental interests must be weighed. Even in a basic search, the court pointed out, federal agents could obtain a great deal of sensitive personal information, using the device's own tools to search for particular words or images.[17]

In fact, the court found that this was precisely what border agents did. The agents asked one of the plaintiffs, a journalist, specific questions about his photos, emails, and contacts. They made notes of the information found on the phone of another plaintiff, a graduate student; they used algorithms to search the work phone of another plaintiff, an engineer at NASA's Jet Propulsion Lab in Houston. The border agents asked yet another plaintiff, who ran a media organization, about the contents of her blog posts and appeared to have reviewed her Facebook friends. They searched the photos of one plaintiff, a nursing student, despite her objections to having male agents see photographs of her and her daughters without their headscarves. One plaintiff's devices were held for fifteen days; another plaintiff saw agents scroll through text messages between her and her lawyer. The border agents also appeared to have tampered with a plaintiff's media files, since the phone was missing video of her daughter's graduation when the government returned it to her.[18]

Based on this record, the court concluded that even the "basic" searches carried out by ICE and CBP were so intrusive that the government must show at least reasonable suspicion (a lower standard than probable cause) prior to conducting these searches at the border. The ruling, in November 2019, marked a significant shift, as it was the first time that a court had ruled suspicionless searches of electronic devices at the US border violated the Constitution. The issue is hardly settled yet; until a case reaches the Supreme Court with a similar ruling, it remains possible that other federal courts will allow comprehensive searches of laptops and cell phones at ports of entry, including for American citizens and green card holders, without requiring even a reasonable suspicion.

WHEN THE PRESIDENT SAYS "SEND HER BACK," WHO EXACTLY DOES HE MEAN?

DNA information raises special concerns, and those concerns are accelerating as the use of at-home DNA tests is on the rise. Ever since the atrocities of World War II, European countries have recognized the ways in which collections of genetic information can be misused. Much of the rest of the world has asked itself the question, "How could ordinary people in Germany have supported the Nazi genocide agenda?" Often, the unstated subtext to that question is, "I wouldn't have done that," or "It couldn't happen here." And yet, in July 2019, President Donald Trump tweeted that four American Congresswomen of color—two of them Muslim, one African American, and one of Puerto Rican descent—should "go back and help fix the totally broken and crime infested places from which they came."[19] When reporters asked White House counselor Kellyanne Conway to explain the president's tweets, she demanded that the reporter, who was Jewish, reveal his ethnicity as a condition of answering his question. As Trump supporters at his rallies echoed his tweet, chanting "send her back" about Minnesota representative Ilhan Omar, a broad swath of the American public was starting to understand how it was possible that a charismatic, autocratic populist leader could whip the populace into a frenzy of us-versus-them outrage and lead people who were easily swayed down a path from ethnic and religious hatred toward violence. It reminded me of a conversation I'd had with a friend in November 2016. A naturalized US citizen and suburban soccer mom who was born in a Latin American country and adopted as a baby by a US family, she was on the verge of tears with worry that the incoming Trump administration would attempt to strip her of her citizenship and deport her to a country she had never known—not primarily because she was naturalized, but because her heritage was Hispanic and her skin color was brown.

It wasn't hard to imagine how DNA information could play an insidious role in a resurgence of racial discrimination in the United States. We're a nation less than a lifetime removed from the official dismantling of Jim Crow—and considerably less than a lifetime from the era when desegregation laws were put into full force and effect. Voting rights are being rolled back as states that had once been under consent decrees for past racial discrimination at the voting booth have emerged from those orders only to institute stringent new ID requirements for voter access, coupled with sweeping purges of voter rolls—measures that, in the past, have been proven to have a racially discriminatory impact.

Only two months before Trump's "go back" tweets, the US Department of Homeland Security had announced that it planned to begin DNA testing of individuals at the border: not just those apprehended for the misdemeanor of crossing illegally or overstaying their visas, but also for asylum seekers hoping to take refuge in the United States.[20] The ostensible purpose of the testing is to ferret out misleading statements of asylum seekers and immigrants who claim to be family but whose relationship is based on affection, or convenience, rather than blood. Immigration hawks contend that migrants coming across the border with Mexico pretend to be families in order to take advantage of rules that often result in family groups being released into surrounding communities more quickly than adults traveling alone. The tests, which would be administered to migrants at ports of entry and at the border—in other words, to those crossing legally and without papers—raise a host of privacy fears. Among them: the tests involve collecting DNA from children who cannot give legal consent; the tests don't take into account meaningful ties of blood and affection that fall short of biological parentage; because the tests are administered as a condition of admission to the country, they are inherently coercive; and by using the same genetic markers that are commonly used for crime-scene testing, migrants who provide DNA samples to prove family relationships for immigration purposes may find their DNA funneled into databases that are used for criminal investigations in the future. Currently, US officials don't have a reference database of DNA of persons who've never been suspected of a crime; this could be a move toward changing that, and it's a move that focuses on immigrants, regardless of legal status.[21] It isn't clear that all of these fears will be realized; DHS officials have said that DNA testing for immigration purposes will not be incorporated into law enforcement reference databases. But the mismatch between biological proof of parentage and legitimate bonds, ranging from adoption to affection, raises opportunities for the kinds of misuse that led Cold War–era officials to doubt the legitimacy of familial claims by Chinese immigrants to the United States.[22] By October 2019, a senior official in DHS announced the administration's continued interest in eliminating birthright citizenship—guaranteed by the Fourteenth Amendment as one of the corrective measures to right the wrongs of slavery—and his belief that this could be accomplished by mere executive order.[23]

That same month, DHS announced what has been described as an "immense expansion" of DNA programs, as it prepared to begin DNA testing of tens of thousands of people who were in immigration custody, and enter those DNA results into national criminal DNA databases. Previously, the FBI's Combined DNA Index System, or CODIS, database consisted primarily of samples from persons arrested, charged, or convicted of serious crimes. This new collection would amount to the creation of a massive reference database of DNA samples from people, including children, who aren't under suspicion of being connected to the types of serious crimes that have been the defining feature of law enforcement DNA databases in the past. [24] The irony, of course, lies in the substantial body of research demonstrating that immigrants commit crimes at lower rates than native-born US citizens. [25]

The concerns about the use of DNA information are real, and creating widespread reference databases of targeted groups raises a host of concerns. At the same time, DNA can provide invaluable evidence to support conviction or exoneration in a criminal case. The same technology raises different privacy issues in different circumstances. On the one hand, the widespread collection being undertaken at the border, for legal and undocumented immigrants, raises serious questions about racial profiling, threshold for collection, and the risks of coercion and challenges of meaningful consent. On the other hand, when police are investigating serious crimes, like the recently solved Golden State Killer case, it isn't clear that it makes sense to require a warrant based on probable cause before police can search a commercially available database, or even collect items from the trash bin that contain the suspect's DNA. Limiting government collection to individualized suspicion would help protect privacy, and restricting searches of information to instances in which a judge has signed a law enforcement warrant isn't a cure-all, or even necessarily an effective approach. It would prevent law enforcement from accessing DNA data for the next Golden State Killer investigation. The wide-ranging examples of government use of DNA illustrate the fact that, if the government is allowed access to DNA data, it's necessary to impose stringent restrictions, overseen by internal authorities as well as by external, independent bodies, to make sure that the information isn't misused. Saying "get a warrant" for every type of personal data that could be misused amounts to taking an easy, intellectually sloppy way out, providing the false appearance of privacy protection while undermining legitimate security goals. The more meaningful, difficult, and necessary policy work has to be in defining new restrictions that will define when, why, and how government can access that data; what the permissible and impermissible uses are; how compliance will be overseen; and what the consequences will be for the government if it mishandles the information. The authors of the Federalist Papers couldn't have imagined the extent to which members of Congress have abandoned their Article I authorities in favor of devotion to their political party. So far, however, the continued vigilance of journalists, citizens, activists—and the judges and politicians willing to place principles before partisanship—has continued to surface wrongdoing when it arises and provide the mechanisms to protest and combat it.

Despite extraordinary overreach at the border, despite threats to persecute political enemies, there are still legitimate and pressing national security and law enforcement challenges, and there are legally and ethically sound reasons why government needs to have data collection and analysis tools. As citizens, we should aim to be as well informed as we can about government use of digital data, and to seek out and critically assess credible news information on these topics. Our goal as a society should be to constructively support legitimate government needs while protecting individuals from overly intrusive, or abusive, data collection and use by the agencies charged with intelligence, law enforcement, and other government functions. We should be on guard for intrusion, but sidestep paranoia. The Reagan-era adage about nuclear disarmament is one that still serves well: "Trust but verify" should be the motto of every American wanting to know how their government is using personal data, and how those uses comport with the requirements of the Constitution and the principles of our nation.

SECTION V

Global Rules in a Connected World: How Other Countries Handle Data

I traveled across eight time zones while writing this book. The process was seamless: the biggest hiccup I faced was when, sitting in a hotel room in Germany, Netflix kept insisting that the Finnish crime drama I was watching should be subtitled in German instead of my usual English. We're long past the days when international trips meant traveler's checks, special-purpose phones, and largely remaining incommunicado until the customs officer stamped our passports on return to our native lands. And it isn't just us: Data travels around the world in nanoseconds. Individuals travel across continents and use their smartphones everywhere they go. The major platform providers are global corporations with data centers in every major region of the world. Email, social media accounts, and apps hardly skip a beat, no matter how many thousands of miles we travel.

Although data crosses international boundaries at the speed of light, the laws that govern data privacy change at every border crossing. A thorough accounting of international data privacy laws would easily fill another book, but three key areas of the world are taking very different approaches from the United States, and wielding significant influence on data privacy within their own territories and around the globe. On one end of the spectrum, the European Union, which recognizes individual privacy as a human right, has enacted a sweeping privacy law that is reshaping data privacy practices everywhere. At the other end of the spectrum, China is implementing the world's most pervasive surveillance system in order to maintain control of its population, and Russia is leveraging the power of digital data to influence the citizens of other nations and to project its power and influence on political processes in countries around the world.

This section takes a high-level look at these three major players in the international data privacy scene.

CHAPTER 19
A BRIEF EUROPEAN (DE-)TOUR

Or Is Being Forgotten Really a Right?

D uring the summer of 2016, I was one of a small number of invited attendees at an international privacy conference being held in Berlin, Germany, in a hotel at the foot of the Brandenburg Gate where the conference rooms overlooked the stone monoliths of the Monument to the Murdered Jews.

The conference attendees hailed from the United States, Europe, Asia, and South America. Among us were current and former data protection authorities and technologists, a representative of the US Department of State, in-house counsels from multinational corporations, academic researchers, accountants, lawyers, and partners in law firms.

Berlin has a gritty energy that refuses to be whitewashed or categorized. The city can't be called beautiful, really: the brutalist, functional, industrial architecture is an ever-present reminder of the cost of rebuilding from war. Other memoranda are prominent throughout the city as well: The remnants of the Berlin Wall that divided East from West Germany throughout the Cold War. The Checkpoint Charlie museum and tourist attraction. The monuments, large and small, spread throughout the Tiergarten commemorating the imprisonment and slaughter of so many people during the Second World War. The still-languishing recon-struction in former Eastern sections of the city, and the massive monuments to Soviet might.

Among the most viscerally powerful exhibits is the indoor-outdoor museum built alongside an intact section of the Berlin Wall. Named the Topography of Terror, the museum sits on the site where, between 1933 and 1945, the "protection squad" of the Nazi party, the SS, and Gestapo had their headquarters. While walking through the exhibits, I was struck with a newly palpable understanding of how European data protection law evolved in the ways that it had, with special consideration to the ways in which demo-graphic data, as well as individualized information, could be used for oppres-sion, persecution, and slaughter.

According to the museum's materials, the most important tasks of Himmler's SS were:

- rigorous surveillance, persecution and elimination of all people regarded by the Nazi state as "opponents" and "enemies"
- creating a "racially pure" Germany, in particular by systematically persecuting and expelling the Jews
- conquering Lebensraum (living space), especially in Eastern Europe, and establishing a political/ethnic "New Order" in Europe based on racist ideology[1]

In many museums, and in many fictional portrayals of Nazi ambitions during World War II, the narrative skips forward to the Holocaust, without the slow, inexorable walk through the surveillance, propaganda, and social control that laid the foundation for so many German citizens to be ready to support a campaign of extermination against others whom they viewed as different from themselves. In this case, however, and appropriate to the methodical and painstaking work of the SS and SD, the Topography lays out the inexorable path to dehumanizing the "other" that was conceived and executed from the building that sat on the same soil where the museum lies today. The museum's materials remind visitors that it was Himmler who created the concentration camp system for removal and slaughter of anyone who was "an enemy of the state and the people," and that those "enemies" included political opponents, Jews, Slavs, gays, the physically and mentally disabled, and many more.[2] These plans were carried out in part through Gestapo surveillance.[3] The tactics of the Gestapo, SS, and later the RHSA put on sickening display the full-embodied capacity of authoritarian surveillance to shape behavior and opinions throughout a nation.

This is the legacy that gave rise to European data protection law, in each of its evolutionary stages since the conclusion of World War II. It's no accident that European data protection law affords the greatest protection to precisely the kinds of demographic information that were so grossly abused as the basis for persecution during the Second World War. The General Data Protection Regulation (GDPR), mentioned earlier in the book, confers additional protections on "special categories" of personal data that include racial or ethnic origin, political opinions, religious or philosophical beliefs, trade union membership, biometric data, health data, and data concerning a person's sex life or sexual orientation.[4]

WHAT EXACTLY DOES EUROPEAN DATA PRIVACY LAW DO?

European law recognizes individual privacy as a fundamental human right, with many principles of privacy law stemming from that foundation—a contrast to the United States, where the Constitution doesn't explicitly recognize a privacy right. Current European data protection law focuses heavily on restricting data collection and allowing "data subjects"—people—to know what data others hold about them and why, and to object where they believe those holdings are inaccurate or improper. The law also aims to foster the growth of a tech industry in Europe, using restrictions on international data transfers outside the European Economic Area (EEA) to bolster data storage and handling—and a range of associated economic activities—inside the EEA. Unlike in the United States, where Congress has authority to legislate on all subjects, the EU is "competent" only on certain matters. The effect of this is that privacy regulations are enacted at the regional, EU level, but each EU member nation retains the right to manage its own policing and national security laws. European law is intentionally extraterritorial, purporting to impose legal obligations on countries with no physical presence in the EU, solely on the basis that they market goods and services to the EU or collect data from residents of the EU.

European law also provides for a "right to be forgotten," a concept largely unrecognized in US law and which raises a number of questions about competing societal interests such as freedom of the press. Perhaps because so many multinational entities are subject to GDPR, European data privacy law principles are being actively considered in the United States. Their influence is evident in the California Consumer Privacy Act that took effect in 2020 and in proposals for federal data privacy legislation being considered by Congress.

IS BEING FORGOTTEN
REALLY A RIGHT?

In 2010, Mario Costeja Gonzalez filed a complaint with the Spanish Data Protection Authority. Gonzalez claimed that when he did web searches for his name, Google's search index served up news articles referring to the decade-old forced sale of his home to satisfy debts. The articles linked by Google were accurate, so he wasn't claiming this was defamation. Rather he claimed that because the sale had happened some years before and the matter was resolved, these old stories were now "irrelevant." He had asked the newspaper to take down its links to the articles, and he asked Google to remove the articles from its indexed search results. When the newspaper and Google declined, the case went to court, and eventually was appealed to Europe's highest court, the European Court of Justice.

The ECJ walked through an analysis of Google's status under European data privacy law, and noted that, under some circumstances, the search engine operator is "obligated to remove links to web pages" that are published by third parties, even when the publication itself is lawful and the articles may remain posted on those pages.[5]

The court's decision rested on the power of search engines that make it possible for anyone with internet access to retrieve information that could touch on almost any area of a person's private life and that wouldn't otherwise have been available—or would, at the least, have been more difficult to find. In the court's view, because search engines make detailed individual profiles "ubiquitously available" and have the potential to so significantly intrude on an individual's privacy rights, the "interference cannot . . . be justified by merely the economic interest which the operator of the engine has in the data processing."[6] The court acknowledged that de-indexing requests had to balance with the public's legitimate interest in having access to information. Over time, the court wrote, information becomes sufficiently stale—"inadequate, irrelevant . . . or excessive in relation to the purposes" for which it was first posted—that the data subject has a valid basis for requesting removal of the link to it.[7] With that decision, the European privacy doctrine of the "right to be forgotten" was enshrined in EU privacy law.

Some subsequent cases have raised questions about how to strike the right balance, like the decision requiring de-indexing an unofficial "blacklist" that included accurate information about a Dutch surgeon who had been subject to medical board discipline for negligence less than five years before—accurate information that prospective patients might want to know.[8] And the law's effectiveness is limited, since de-indexing is required only for search results displayed within Europe, meaning that the information can still easily be found online. Nonetheless, by 2019, Google had fielded over 800,000 requests and agreed to roughly 43 percent of them, leading them to de-index some 3.3 million links.[9]

The notion that an individual should have the right to demand de-indexing information about them that is accurate but that might be embarrassing or unflattering runs contrary to most practices, and many laws, regarding public records, press freedoms, individual freedom of speech, and public safety protections in the United States. Arrest records, convictions, foreclosures, bankruptcies, divorces, and many other records that may be unflattering are readily available online in the United States, with no movement yet toward widespread protections that would allow individuals a right to have that information taken offline or de-indexed from internet searches.

There is one area, however, where an analogy exists: the ban-the-box movements that are prompting state and local legislators to ban employers from asking questions about criminal arrests (or sometimes criminal convictions) on job applications. The rationale is that these questions often aren't relevant to the particular job, and so legislatures try to tailor them accordingly. Some laws prohibit asking about arrests but allow asking about convictions. Some laws prohibit these questions on the initial application, but let employers ask about criminal history before deciding whether to make a final job offer. And many states limit criminal history inquiries based on job categories, allowing the questions for positions that involve working with children or the elderly, for public safety jobs, and other job categories where a criminal record might be particularly relevant. As of 2019, nineteen US states had enacted some form of a ban-the-box law, and a number of major corporations, including Walmart and Target, had adopted ban-the-box policies in the hiring process.

Perhaps these ban-the-box statutes and policies signal a growing sense in the United States that, while a right to be forgotten may not be absolute, society ought, at least, to give individuals the opportunity to put their past behind them.

IS EUROPEAN PRIVACY LAW REALLY MORE COMPREHENSIVE THAN PRIVACY LAW IN THE UNITED STATES?

European privacy regulators, and many privacy advocates, argue that European privacy law is stricter and more comprehensive than in the United

States, and the official EU position is that the United States doesn't provide "adequate" privacy protections for personal data. But it isn't clear that either of these assertions is entirely true. Europeans (and others) have long criticized US data privacy laws on the grounds that, because they are generally sector-specific, they are inherently narrowly defined. There's some truth to that, and this book has described a number of areas of growing disconnect, such as the gap between situations where health-related information in the United States is and isn't protected by HIPAA. Other examples include financial services, in which the Graham-Leach-Bliley Act (GLBA) requires banks and other regulated entities to maintain the confidentiality of financial transaction information, but doesn't restrict banks' use of personal data for purposes like marketing. In the higher education context, the Family Educational Rights and Privacy Act (FERPA) requires most colleges and universities to obtain consent from students over the age of eighteen before sharing academic information with anyone outside the university, including the student's parents. But schools are free to use personal data for activities like marketing to prospective applicants and identifying potential donors. So there are clearly gaps. By and large, however, GDPR doesn't close those gaps; it simply applies a one-size-fits-all approach to all uses of personal data, no matter which industry sector might be using it.

The other primary criticism that European privacy regulators have directed at the United States is the claim that the US is overly aggressive in its intelligence collection and national security programs. Ever since the 1980s, the European Union has looked at the interrelationships between personal data, surveillance technology, and the United States with concern—but not with much indication of self-reflection on the government surveillance activities of European nations. Here, a brief history lesson helps.

In 1980, the Organization for Economic Cooperation and Development (OECD) adopted Article 108, the first comprehensive statement of multinational data privacy principles. These principles were modeled closely on the Fair Information Privacy Principles, or FIPPs, developed by the US Department of Education in the 1970s for providing a privacy protection framework for computerized information.[10] At the time, European nations were actively looking for ways to negotiate together on matters of mutual importance in the region, particularly with respect to economic development and prosperity. Europe had been hit harder than the United States by the global recession of the early 1980s, and European countries were suffering through record levels of unemployment.[11] By 1993, when the Maastricht Treaty that created the EU took effect, European nations were increasingly uneasy over the technological dominance of the United States in the telecommunications and infrastructure sectors. They were keenly aware that Europe's technology sector was lagging far behind and were concerned about the economic impact of America's upper hand in this centrally important economic sphere: after all, the United States was building most of the hardware backbone that global telecommunications would run across.

While the EU was struggling to recover from recession and come to terms with its new regional identity under the Maastricht Treaty, European lawmakers began to pay increased attention to a long-simmering set of murky news stories about ECHELON—a poorly defined but sinister-sounding program of electronic surveillance allegedly carried out by the US Intelligence Community. The term had been used in occasional articles dating back to the 1970s but took on renewed interest in the late 1980s and 1990s, partly as a result of books published by journalists in the United States and New Zealand. There were other actors pushing the ECHELON story, though. According to Sir Robert Hannigan, the former director of the United Kingdom's Government Communications Headquarters intelligence agency, the uptick in European outrage over ECHELON was the result of a Russian information operations campaign, deliberately planting information that would inflame European distrust of the United States.[12] The campaign worked. Never mind that the allegations were that the ECHELON program was operated jointly by the United States and its Five Eyes partners: Australia, Canada, New Zealand, and the United Kingdom. From a European perspective, the heat was now on the United States. These reports provided European lawmakers, already inclined toward economic protectionism, with the ammunition they needed to insist that there be restrictions on the transfer of personal data from European countries to the

United States where, it was asserted, the laws did not adequately protect the privacy of personal data.

The approach has been consistent over time: in each iteration of European data protection law and the accompanying treaties to allow European data to be transferred to the United States, the EU has pointed to both privacy and economic reasons for the regulations. In the most recent law, the GDPR, the first recital points to the central nature of privacy as a human right. The second recital notes that:

> This Regulation is intended to contribute to the accomplishment of an area of freedom, security and justice and of an economic union, to economic and social progress, to the strengthening and the convergence of the economies within the internal market, and to the well-being of natural persons.[13]

In other words, when it comes to regulating the free flow of data, the priorities that count are fostering "economic and social progress" of the EU, and to "strengthen and converge" the economies within the internal European market.[14]

Perhaps because the conversation has so often been couched in terms of whether US data privacy law is "adequate" when compared with EU law, it often loses sight of the economic protection dimensions. It also often ignores the fact that European data privacy law takes little meaningful account of the data collection or processing activities carried out by its member states for national security purposes. This is a critically important difference in the two legal frameworks. In the United States' sector-specific approach, regulations relating to law enforcement and national security must consider the privacy and civil liberties that stem from a number of provisions of the Constitution and common law. In the EU, however, the GDPR's "comprehensive" data protection law explicitly exempts all of its own member states from any potential restrictions or oversight relating to intelligence activities. The reason for this is the EU can only legislate in areas in which is "competent," and under the European charter, the Union may not legislate on issues relating to law enforcement, national security, or public safety. Each of those areas is reserved to the member states.

So, do European countries carry out wiretaps of their own residents and citizens for law enforcement and national security purposes? You bet they do. Do they collect emails—metadata and content—from the internet? Absolutely. Do they conduct computer network operations that include gaining surreptitious access to the digital devices and accounts of individuals and groups who are suspected of being a threat to public safety? Most definitely. The difference between the US and the EU in this regard is that the United States has in place far stronger safeguards and privacy protection mechanisms than most European countries do. In many European countries, judicial review of surveillance requests either isn't necessary or the judiciary isn't a structurally independent branch of government from the law enforcement or intelligence agencies that are requesting the approval. In these countries, the legislative body is inherently formed from the same parties that won election to run the executive branch. Consequently, parliamentary bodies are more likely to endorse whatever surveillance laws their party's leadership wants to enact. Meanwhile, Congress's fractious relationship with the White House and differences in party control of those bodies often means that Congress serves as a significant stumbling block for administrations that may want to enact privacy-intrusive surveillance laws. Similarly, parliamentary oversight is often less meaningful than congressional oversight in the United States, where Congress and the executive branch are separate and independent divisions of government.

Further compounding the scope of potential privacy intrusions from European national security activities, many European countries have state ownership or state investment in their telecommunications infrastructure, meaning that the distinction between private- and public-sector roles in safeguarding, and facilitating collection of, electronic communications is not as clearly drawn as it is in the United States. A state-owned telecom provider is not, for example, going to sue a European government protesting an order to turn over communications that may be relevant to a law enforcement investigation. In the United States, by contrast, those kinds of lawsuits have happened, and will likely continue.

MORE IMPORTANT, IS EUROPEAN LAW MORE EFFECTIVE THAN PRIVACY LAW IN THE US?

Despite the intention to incorporate the lessons of history into data protection law, it isn't clear that the GDPR's approach is well calibrated to achieve that goal. Much of Nazi oppression and persecution was based on publicly available, or publicly observable, information about a person's status: that an Aryan woman was in a romantic relationship with a Polish man; that a family attended synagogue, or were members of the Jehovah's Witnesses; that workers belonged to a trade union; that a child was blind, or an adult was a wheelchair user or had some other physical limitation; that particular individuals had been handing out leaflets on behalf of the Communist Party. Today, the same kinds of information are more available than ever: digital profiles the Nazis couldn't have dreamed of are available through a variety of public, private, and volunteered sources. In order to protect privacy, GDPR relies heavily on limiting data collection. But it isn't clear with what effectiveness, or to what end. In some cases, the data is publicly available; restricting its collection doesn't make the information disappear.[15] In other cases, the law is being used to restrict processing that is relatively benign.[16] And since GDPR doesn't apply to national security and law enforcement activities, it doesn't protect individuals in any meaningful way from surveillance programs carried out by EU member states, no matter how intrusive or oppressive they may be. For all these reasons, GDPR may not be the best approach to protect people today from the kinds of persecutions that took place in the past.

With these challenges in mind, a growing number of thought leaders are considering whether currently favored strategies of limiting data collection and relying on notice-and-consent have outlived their purpose. In this evolving school of thought, perhaps the better approach is to accept that data will be created and collected in nearly limitless volumes; that once it exists, it's virtually impossible to eradicate it; and that purely computational or organizational steps like filtering, indexing, cataloguing, scanning, tagging, and even retaining personal data don't always harm privacy rights, if there are enough protections in place to guard against substantive harms from the data's misuse. In other words, this newer approach says we should focus first on identifying the harm we want to prevent, and then choose from a flexible menu of mechanisms that could be used to protect against that harm. Those tools might include limiting collection or requiring notice-and-consent, but the mechanisms would be merely a means, not an end in and of themselves. With this more holistic approach, there might be fewer cases that elevate preference for a particular privacy mechanism above common sense, and we might be better able to prevent substantive harms by placing greater priority on the ends than the means.

CHAPTER 20
TOTAL(ITARIAN) SURVEILLANCE

How the Other Half Lives

t's a free country. I can say what I want!"

These words ring out across every playground in America, closely followed by another key principle derived from our Constitution: *"You're not the boss of me!"*

Throughout the United States, children grow up knowing, and demanding, their inalienable right to free expression. Although the commitment to free speech has frequently been strained by the political strife of each specific era in American history, that right of free expression remains jealously guarded by American legal, cultural, and political traditions. The First Amendment to the Constitution, along with the Fourth and Fifth Amendments, is one of the nation's most powerful defenses against sustained overreach by the government in surveillance activities that could impact US persons. That isn't to say that there haven't been, or aren't, legitimate concerns about US surveillance and intelligence activities—some of the ugliest chapters in America's past intelligence operations are described in earlier chapters of this book. Nonetheless, for every US surveillance program that overstepped the bounds of what was lawful, authorized, or ethical, other nations throughout the world have gone much further in harvesting and using data about their own citizens and those of other nations.

Think tank reports have shown steady year-after-year declines in internet access, with 36 percent of the world's major countries imposing stringent restrictions on their residents' ability to access information online.[1] In nineteen countries across the Middle East, Africa, Eastern Europe, and Asia, internet access is so heavily restricted and censored that residents effectively lack free access.[2] For example, although Bahrain has one of the highest rates of internet access in the region, all ISPs are under government control; the government has responded to civil unrest with internet takedowns; all ISPs must block access to sites based on a government-mandated filtering list; and the government has used that authority to block sites as mainstream as Al Jazeera and apps as widely used for secure communications as Telegram.[3] In Ethiopia, the government controls the telecommunications infrastructure; internet access is limited; internet shutdowns are ordered during periods of political action and civil unrest; and social media and file sharing sites like Facebook, Twitter, Dropbox, and WhatsApp are routinely blocked to suppress political speech.[4]

While most repressive regimes rely on sophisticated blocking technologies to restrict access to web content and services, Cuba has simply made internet connectivity, and access to internet-capable devices, largely unavailable, in order to effectuate content censorship.[5] Vietnam has forced the takedown of hundreds of social media accounts and thousands of videos that were critical of the state.[6] In the UAE, the state systematically blocks access to websites relating to pornography, gambling, political content, religious content, human rights, and LGBTQ issues, and has criminalized many forms of digital free expression.[7] The list could go on, a depressing litany of the ways in which authoritarian regimes combine internet surveillance with other forms of data collection and social control to strangle political opposition, limit human inquiry and expression, and interfere with rights of association, religion, and personal dignity.

Some of these actions—such as requiring all internet traffic to be stored on state servers or run through state filters—are inherently privacy intrusive. Others, like blocking access to information or imposing penalties for expression, are actions in which data privacy rights intersect with other intrinsic human rights and needs that are oppressed, sometimes violently, by the state. Many of these examples are illuminating both as cautionary tales about how intrusive data surveillance can be, and also as models that demonstrate which of the laws and norms of Western democracies currently stand in the way of a further spread of these total-surveillance approaches into other countries.

THE CYBER CLUB

The list of highly cyber-capable nations isn't long, roughly similar to the number of nuclear powers in the world. Although not identical, the membership rosters for the two clubs looks much the same: a mix of US allies and adversaries, of nations who espouse the values of liberal democracy as well as those who look at digital data and see cyber capabilities as another tool set for authoritarian repression. Like most "best of" lists, the rank and order lies in the eyes of the beholder, but almost any observer of cyber operations would include China, Russia, Iran, Israel, and the United Kingdom as among the most cyber capable.[8] The US government's spy chiefs have warned that the nation's primary cyber adversaries are China, Russia, North Korea, and Iran.[9] Unlike nuclear powers, however, cyber capabilities come in different forms, and there are a number of countries who are highly skilled in some areas of cyber operations, and less focused or less formidable in others. When it comes to cyber defense—the ability to detect and prevent cyber intrusions and attack—perhaps it's not surprising that some of the nations with the most robust programs are countries who have experienced, or are at greatest risk of, cyber aggression from world powers in their region. Estonia and Norway make the best-of-cyber-defense list; so do Singapore and South Korea.[10] And, of course, plenty of cyber capability rests with non-state actors as well.[11]

For the purposes of this book's focus on the ways that digital personal information can be used and misused, the most salient features of a national cyber program are a country's ability and willingness to acquire personal data about individuals and use that data to advance the anti-democratic goals of an authoritarian regime. Those goals typically come in two forms: internal surveillance of a regime's own people, or the use of personal information for targeted messaging, propaganda, and manipulation of the opinions of individuals in targeted regions around the world. Looked at through this dual-focused lens of cyber-enabled internal surveillance and cyber-enabled global influence operations, there's no doubt that China and Russia top the list.

These two digital rivals, China and Russia, have managed their branding and market segmentation well. Both countries have formidable, full-spectrum cyber capabilities; both have flexed that muscle in ways that go well beyond traditional intelligence gathering; and both have leveraged digital technology as a means to enhance or consolidate their political power. But they've approached these strategic goals differently, and both serve as examples of the dangerous consequences that can result when autocratic governments have unbridled access to digital information, and use it in ways that are divorced from, or indifferent to, the commitment to civil liberties and individual freedoms that forms the bedrock of Western liberal · democracy.[12]

RUSSIA

If there's an art to optimizing the use of personal information for propaganda and information warfare, Russia has mastered the discipline.

> Russian political culture sees information as a threat—a weapon that you can use or that can be used against you, and that should be controlled at all costs. The Kremlin doesn't see surveillance, domestic media policy, cybersecurity, and internet governance as separate issues. Rather, they are deeply connected to misinformation campaigns abroad, and are used strategically to achieve geopolitical goals.[13]

WESTERN DEMOCRACIES AND AUTOCRATIC POWERS: IS THE POT CALLING THE KETTLE BLACK?

Observers can, and do, regularly point to the risk of overreach in countries like the US and UK; but each of those nations has a strong tradition of free speech coupled with scrutiny of government operations and robust debate about what government law and policy should be. These features of liberal democracies aren't enough, by themselves, to guarantee that intrusive government surveillance can't happen there, for a variety of reasons. First, reasonable people will continue to disagree about where to strike the balance between individual liberty and collective security; these policy choices, no matter how transparent, will sometimes result in actions that some citizens view as abhorrent overreach. Second, the rapid rate of digital change will sometimes make it hard to predict the beneficial and negative consequences from new technologies. Third, judges and legislators will continue to be hard-pressed to apply old laws to new technologies, or make new laws at a pace that's rapid enough to keep pace with each new challenge as it arises. And fourth, liberal democracies, like other forms of government, can be hijacked by corrupt and self-serving leaders who will attempt to use the apparatus of state if their efforts are left unchecked. All that said, the constitutions, political structures, press freedoms, and cultural values of liberal democratic nations provide a basis for thoughtfully assessing, with a skeptical eye, what the privacy and liberty impacts of each evolution of digital technology might be.

Russia has a long and well-documented history of using information operations to achieve both domestic political and foreign policy goals. Inside the country, Russia's secret police—the KGB during the Cold War and the FSB since the fall of the Soviet Union—has eavesdropped on citizens, read their mail, bugged their homes and workplaces with listening devices, and steadily amassed files of information about any statements or activities or leanings that could be construed as detrimental to the state. Russian authorities have encouraged citizens to snitch on each other, to report their neighbors' bad or questionable or unsavory behavior. The deeply rooted inclination toward tightfisted state control and intrusive investigation of citizens' actions, speech, and opinions dates back to the Stalinist purges, with only a brief interruption after the fall of the Soviet Union and before current Russian leader Vladimir Putin began consolidating power in order to restore an autocratic, corrupt rule in Russia.

Russia has truly excelled in the area of information operations: of spreading lies, disinformation, and misinformation, of engaging in propaganda campaigns that are so multifaceted and multi-pronged that it becomes hard for individuals to know where factual accuracy lies or to differentiate the true from the false. Russia's interference in foreign elections—in the 2016 Brexit vote in the United Kingdom, the 2016 US presidential election, and subsequent elections in France, Germany, and elsewhere—is well documented and has become increasingly aggressive worldwide. The success of these campaigns depends on using detailed personal information to manipulate voters in the targeted nations.

Russia's surveillance activities have by no means been limited to external operations, however. In addition to the use of personal data to sow global political instability and exacerbate discord within and among Western democracies, Russia's internal surveillance activities have translated long-standing Soviet priorities and predilections into a modern context that leverages modern technology. For example, prior to the Sochi Olympics, the US State Department warned all US travelers to Russia that their communications would be monitored throughout their stay by the System of Operative-Investigative Measures (SORM), Russia's national system for government collection of communications data. SORM has been in place since the 1990s, and is used aggressively by Russian authorities to capture meta-data about internal Russian communications and to effectuate the millions of wiretaps that have been authorized under Russian law. The State Department's travel warning included the following advice:

> Consider traveling with "clean" electronic devices—if you do not need the device, do not take it. Otherwise, essential devices should have all personal identifying information and sensitive files removed or "sanitized."

The advice went on: Keep Wi-Fi disabled at times. Don't connect to the internet at coffee shops, hotels, airports, or other venues. Change all your passwords before and after your trip. Remove smartphone batteries so that your phone can't be remotely activated. If you have to use a mobile phone, make it a burner phone purchased locally with cash. "Sanitize" your conversations.[14] This wasn't advice for key government personnel; this was advice for ordinary tourists, because the data privacy surveillance risks in Russia are so high.

The FSB operates the SORM with a degree of independence and lack of any meaningful oversight that, from a Western democratic perspective, is both striking and alarming. The FSB has to obtain a court order to tap a phone line—but they aren't required to show the order to the telecommunications provider. And the way that the legal and technical architecture for SORM is constructed effectively ensures there will be no external oversight: the collection equipment is installed directly by the FSB, data collected from SORM is accessed directly inside FSB offices around the country, and the communication providers have no access to the SORM equipment themselves. Citizens have no way to know if they're being monitored; the provider has no way to know whether a court order was actually issued; no one at the provider needs to assist the government in obtaining the data; there is no after-the-fact review by anyone outside the government agency carrying out the surveillance; and—although the FSB has primary responsibility for SORM—there are at least seven government agencies who have authority to intercept communications using this technology.[15]

The surveillance state has been strengthened with the passage of new laws designed to limit anonymity in digital communications. In 2014, Putin disparaged the internet as a "special CIA project," passed a law allowing the state to block websites, and enacted new restrictions requiring all bloggers with more than three thousand followers to register their blog with the government.[16] In 2016, Russia passed the "Yarovaya Law," named after its key author, Irina Yarovaya, and referred to as a "Big Brother law" by critics, which requires telecommunications companies in Russia to store copies of all text messages, phone calls, and emails, available for government access, for a minimum of six months. Although couched as a privacy-protective measure, attempting to piggyback on the international data transfer restrictions of EU law, there's simply no credibility to any claim that the law advances privacy goals. The law also allows copies of the data to be exported from Russia. In other words, the data localization requirement is intended to ensure Russian government access; so long as that's achieved, it's of little importance from a government policy perspective who else might have access to the information. In 2017, Russia passed laws outlawing VPNs that could carry private, encrypted communications and banned the use of TOR, an anonymization service that disguises the true IP address associated with a user's web traffic. These new laws made clear that the goal wasn't to protect individual privacy, but the opposite: to ensure that government surveillance could intrude on virtually all communications in Russia, effectively making privacy against the law.

CHINA

Meanwhile, if there's an art to constructing a totalitarian surveillance state, China has mastered it.

China has been in the business of censorship and electronic surveillance for years but has been making headlines in unprecedented ways with the build-out of its surveillance state in the 2010s. The Chinese government is leveraging every modern form of digital information in carrying out these goals: internet search and communications (via email, chat, direct message, and on web forums); surveillance cameras coupled with facial recognition databases; social media platforms, including accounts used by government officials, to polling, interaction with Chinese citizens, and other forms of what has been called "authoritarian deliberation"; encouraging individuals to report suspicious undesirable or antisocial behavior (such as jaywalking) by their neighbors; and the use of massive databases and algorithms to correlate and combine all this data as part of a sweeping program to create a system of "social credit scores" for people living in China (more about this later).[17] In recent years, China has even passed "cybersecurity" laws that require all data about Chinese nationals to be physically stored in China—ostensibly for data protection, but with the obvious benefit, from a governmental perspective, of requiring "network operators" to allow Chinese government officials to have full access to data and to provide "technical support" to the government on request.[18]

To put things in context, China has been censoring the internet since the internet first arrived in the nation. When China began allowing its citizens to access the internet in the 1990s, the move was initially heralded as a positive one, a hallmark of China's new openness to engage with the global community as its role as an economic power on the world stage grew. As part of China's "Great Firewall" blocking access to Western news media, the government jealously guards the scope of information that Chinese residents are permitted to see. China's leading social media platform, Weibo, functions much like Twitter, and had an estimated 500 million registered users at the beginning of 2020—a userbase within a single country that many companies would relish having worldwide. On Weibo, however, government censorship is common, with some observers estimating that the government might have as many as a million people working as censors, looking for sensitive, inflammatory, incendiary, divisive, or otherwise unpalatable content. Their role is wide-reaching, as is their interpretation of what information might be considered controversial or inconsistent with the goals of the state. Some of the examples reported by Western news organizations included memes that would be considered tame by the standards of Twitter or Instagram: a line of rubber ducks mimicking tanks; a hand of playing cards whose numbers spelled out the date of the massacre in Tiananmen Square; even posts showing photos of candles, for fear they might be interpreted as lighting a candle in memory of the protesters who perished and of the protest movement that was crushed that day.[19]

Thirty years after Tiananmen Square, protests erupted in Hong Kong during the summer of 2019 over a new law that would allow extradition of Hong Kong residents to mainland China. This new generation of protesters turned out to be savvy natives of the digital ecosystem. Hong Kong's metro system slowed to a crawl as protesters stood in line to purchase tickets for cash, eschewing the automated metro fare cards that would speed their progress through the turnstiles but which would also record their movements from one station to another. Protesters used secure messaging apps to communicate with each other and deleted social media posts that—even in relatively tolerant Hong Kong—could bring unwanted attention from the government. They turned off location services on their phones; they wore masks and goggles, partly to protect their respiratory systems from tear gas, and also to foil facial recognition systems connected to the ubiquitous surveillance cameras.[20] When face masks were outlawed, they created projection systems worn like hats that distorted the light falling on their faces to render FRT ineffective, and even posted YouTube videos with hairstyles designed to cover key facial features, again to defeat FRT. There was nothing about these digital self-defense measures that would, if needed, protect a protester from tanks rolling down the street. But in the decades since the Tiananmen protests, it had become clear that digital surveillance tools were more pervasive, more powerful weapons in the Chinese government's war against dissidence than any heavy artillery could be.

SPREADING SURVEILLANCE

China has become infamous for the scope of its surveillance state, and the detailed data it collects is regularly used in ways that are chilling, from the oppression and incarceration of the Uighur minority group, to the suppression of protests in Beijing, to establishing a "social credit score" for all Chinese residents, which incorporates personal information from the ridiculous to the sublime: where someone works, what grades they earned in school, whether or not they were caught jaywalking by one of the ubiquitous facial recognition cameras, and whether their neighbors have lodged a complaint against them for some kind of undesirable behavior. This last bit—encouraging, or requiring, individuals to inform the government when their neighbors misbehave—bears the hallmarks of authoritarian governments everywhere.

It also goes to the heart of one of the key dimensions of privacy: the ability to have limited conversations, to share certain information with some people but not others. If no one can be trusted, if even close friends or relatives could turn "state's evidence," then the chilling effect on social discourse becomes frigid. The stakes involved in questioning the status quo become intolerably high. Any dissent or criticism of the state is quelled before it can be spoken aloud, and it quickly becomes impossible to know, or even guess, whether there may be other like-minded individuals who share similar criticisms of the state. These techniques—fomenting fear, suppressing dissent, forbidding free speech, encouraging neighbors to inform on each other—are time-worn tactics implemented by authoritarian governments and occupying empires around the world and across many centuries. But today's technologies create an almost infinite range of new means for an authoritarian-leaning state to carry out this kind of oppression, suppression, and control of its populace. The current examples of Russia and China demonstrate all too effectively how real this risk is.

The threat of authoritarian use of surveillance technology may be spreading. Venezuela sent government officials to talk with China about the surveillance technology embedded in China's national identity card. China has exported facial recognition technology to regimes in Singapore, the United Arab Emirates, Zimbabwe, and Malaysia. Russian firms have exported surveillance technology to Kazakhstan, Tajikistan, Turkmenistan, and Uzbekistan. Of course, US-developed surveillance technology can be exported and sold as well.[21] But the United States has for years been a party to a set of international treaties under which Western democracies agreed to limit the transfer or sale of surveillance technology to autocratic regimes.[22]

It isn't just the spread of surveillance technology, though, that should cause us to worry. It's the spread of authoritarian ideology, and the use of surveillance technology as a means of political repression and cultural control in countries around the world, that should give us pause. The First Amendment rights to freedom of religion, expression, association, and the press are among the most cherished freedoms America's Constitution provides; they're central to our idea of what it means to be an American. Yet our systems aren't incorruptible, and may not withstand a sustained assault on democratic values if that assault comes from within.

The most dangerous exports from countries like China and Russia may not be toys covered in lead paint, or plastic toys infused with poisons, or loosely secured nuclear material from former Soviet bases that could make it into the hands of enterprising terrorists looking to create a dirty bomb. Rather, it may be that the normalization of surveillance is the greatest threat to democracy around the world.

SECTION VI

Pandora's Box: Data's Dangers, and Finding Hope at the Bottom of the Box

As we head deeper into the twenty-first century, it's becoming increasingly clear that early ideas about data privacy underestimated the mischief that's possible with the misuse of information. As noted in previous chapters, early data breach laws in the United States focused on consumer privacy and identity theft. States began requiring companies to notify consumers if hackers or malicious insiders or other unauthorized persons gained access to personal data such as Social Security and credit card numbers that criminals could use to set up fake accounts or to make unauthorized purchases in a consumer's name. Running in a parallel track, US and European data privacy laws have frequently focused on notice-and-consent. One aim of this book has been to make the case that privacy is about more than data breaches; that personal data is about more than credit card numbers; that notice must be meaningful and choices must be real; and that, as important as it is to understand how data is created and collected about us,

261

it's even more important to consider the ways that our data is used and the myriad impacts those uses can have on our lives.

For all of us as individuals, the right to have some measure of control over who knows what about us—to reveal some information to some people but not to all, and to keep some information entirely to ourselves—is essential to our sense of personhood and dignity. It's vital to our ability to have deep and trusting relationships with other people, for if we can't confide in a loved one or limit how far our communications travel, then we hardly dare speak what's on our minds. And if we can't speak without fear of repercussion, then we'll hardly be able to formulate our ideas, or to test them against the perspectives of others who might see the world differently. The ability to move anonymously through the world, to carry out our daily living without being identified, recorded, and tracked at every step, is crucial to our sense of autonomy and to a free society. The ability to rise above the mistakes of our past rests in part on the ability for our missteps to fade appropriately into the past, and our ability to rise to our maximum potential—to pursue life, liberty, and happiness—depends in part on our ability to pursue risks, challenges, and opportunities unfettered by some predetermined notion, established by the output of a behavioral prediction algorithm, of what we are capable of and what we are likely to do.

The fallout of recent international election cycles has made clear that it isn't just who has access to information about us that matters; it's also the way that our personal data is used to manipulate us, to undermine our sense of personhood and autonomy, to exploit societal fault lines, and to change our minds.

It isn't just who has access to information about us that matters; it's also the way that our personal data is used to manipulate us, to undermine our sense of personhood and autonomy, to exploit societal fault lines, and to change our minds.

It's one thing to articulate the ways that privacy matters and identify the trends in modern life that threaten to undermine it. It's another challenge altogether to discern what to do about it, given the sometimes conflicting forces at play. Each of these critical notions of privacy clashes, at times, with another important set of principles, and with an inexorable truth. When it comes to competing principles, any sensible approaches to privacy must recognize that our individual rights need to be balanced with other important dimensions of social good, including national security, law enforcement, and public health research—and even more economically oriented social goods, like technological innovation and local and global commerce. The inexorable truth is that data-driven technologies continue to grow, and it would be unrealistic to think that we can put Pandora back into her box. As technology expands, we will continue to generate new kinds of data and in new ways that weren't possible before; after all, a mere decade ago, who would have thought that smart toilets would diagnose our health from our feces, or that social media likes would generate detailed personality assessments, or that our cell phones would be carrying out precision geolocation on our every step? Governments can pass regulations, but protecting privacy by simply limiting data collection is no longer a realistic answer. Most of this data generation and collection is happening because the private sector sees business opportunities to create new products and services that offer some combination of convenience, entertainment, social benefits, and profit. Setting aside the merits of any particular technology, the fact is that most consumers welcome many of these new technologies, even if they have reservations about whether to personally adopt any given one.

The fact that market forces drive much of the new data creation and collection doesn't, however, mean that companies should be let off the hook. The private-sector entities whose business model lays at the center of these conundrums must be called to account when their personal data practices overreach and, if need be, should be compelled by law or regulation to behave as responsible world citizens, accountable for the negative consequences of the corporate data exploitation that they profit so handsomely from. At the same time, it would be shortsighted to take a paternalistic approach in which government regulation so tightly limits the creation and use of new technologies that innovation—which, on the whole, is welcomed by consumers—becomes stifled or grinds to a halt.

We've explored how the private sector, individuals, and governments interact when it comes to personal data and examined the fabric of privacy and the related vital rights and interests underpinning these interactions. This final section looks at some of the future dangers of data-driven manipulation and offers some thoughts on ways in which law and policy can adapt to the rapidly changing data environment.

CHAPTER 21
QUANTUM POLICY

Or How a New Approach to Law and Policy
Could Give Cyber Privacy a Fighting Chance

The unsolved equations arising at the intersection of technology, privacy, and law have brought us to a turning point similar to the revolution in scientific thinking that has taken place over the past fifty years. For centuries, scientists believed that the laws of physics described by Isaac Newton were adequate to explain the workings of the universe. For most purposes they still are: we observe gravity, inertia, and similar forces and agree that Newton's three laws make sense. But at the edges of our understanding—in the subatomic scale and across the vastness of the universe—Newtonian physics broke down. At the scale of things that were unfathomably large and immeasurably small, something in those laws didn't hold true. The scientific evidence we were collecting made clear that Newton's laws were inadequate to explain new scientific questions society was facing. At the intersection of law and technology today, we're facing a similar evolution. It's now possible to collect data that is so granular—subatomic particles of information, if you will—on such a massive, grand, and continuous scale—data collection and analysis that matches the scale of the universe—that our traditional approaches to law and policy struggle to make sense of what these advances mean for privacy and technology, and leave real doubts about whether law and policy can keep up. In the face of these challenges, we need a new approach, something I think of as "quantum policy."

Quantum physics asks us to believe two apparently contradictory things simultaneously: that light can be both a particle and a wave; that bits in a computer can register both one and zero at the same time. In the days of Newtonian physics, propositions like these would have felt like something out of *Alice in Wonderland*. Today, though, we know that Newtonian physics can't explain subatomic behavior or the cosmos. At the scale of the very large and infinitesimally small, we needed a new way of thinking, a realization that led to the development of quantum theory. Indeed, quantum physics emerged because the laws of physics we had relied on before were no longer sufficient to explain things we could observe about the way the universe worked. For the early thinkers, especially, it must have been a daunting task.

We're at a similar turning point when it comes to applying traditional law and policy approaches to the questions raised by algorithms, big data, and individual privacy law.

What does a Newtonian approach to law and policy look like? It's one in which we're comfortable with gradual, incremental change, where we continue to rely on the slow accretion of precedent, content with having critical legal issues decided by disparate cases that take years to wend their way through multiple jurisdictions before arriving at a critical mass of new law that's derived through an organic evolution. It means continuing to require that there be a case or controversy; refusing to allow courts to offer advisory opinions. It means that once a precedent has been established, its inertia becomes almost insurmountable. Each of these attributes of law provides the benefits of caution and seeming stability. But they stand in direct contradiction to modern technology development, where beta versions and user acceptance testing and minimum viable products are the watchwords of the day, where failing fast in order to support rapid improvement in iterations is the key driver behind the technology growth that has become an engine for economic and social change, both for good and for ill.

It isn't possible, of course, to map a precise one-for-one comparison between the evolution of hard science and law and policy. Instead, this chapter offers up quantum physics as a conceptual analogy for the ways in which we might approach the challenge of modernizing law and policy at a time when technological advances continue to surge, and whether when looking at vanishingly small bits of data or the algorithms running across massive data collections on cloud platforms, we need to consider what our new ethics, policy, and law for privacy and technology will be.

SHIFTING THE WAY WE THINK

As a first step toward a theory of quantum policy, we need to shift the way that we think. Everything we thought we already knew remains true in many areas of life—apples still fall from trees, objects set in motion will continue to stay in motion until a countervailing force acts upon them, and traditional ideas of privacy still apply to many of our everyday activities. But as with quantum physics, when we arrive at the realm of the vast or the very small, we need to approach data privacy problems with fresh eyes, willing to consider testing new hypotheses and advancing new theories in order to explain and address the phenomena around us. The list below isn't comprehensive, but it provides a starting place for the change in mind-set that will be necessary in order to maintain meaningful privacy in an upcoming future as digital data becomes ever more enmeshed in the fabric of our everyday lives.

DEVELOPING DEFINITIONS OF PRIVACY THAT REFLECT DIGITAL REALITY

To tackle complex issues raised by our digital ecosystem, we need to develop more, better, and clearer definitions of privacy in the United States and around the world. Right now, we ask a single word, "privacy," to represent too many flavors of related things—we need a taxonomy that allows us to speak with increased nuance and greater clarity about the broad range of rights and interests we mean when we currently use that overburdened word. This new taxonomy needs to include concepts of autonomy, personhood, self-determination, and the ability to have intimate relationships and make limited disclosures of information. It needs to include the ability to move anonymously through the world, to live without having our every action recorded for someone else's analysis and for posterity. It may need to accommodate a great deal more. And, of course, it needs to include the right to be left alone.

We need this more comprehensive taxonomy of privacy types and competing interests in part so that we can be specific about what exactly we're trying to protect in any given circumstance, and why. Clearer, more granular definitions will help us more effectively debate the kinds of privacy that we think are most important to protect, and how the rights of the individual should be weighed against other, equally important, societal interests.

TREATING DIGITAL DATA AS MULTIDIMENSIONAL

If we're going to develop privacy-related law, policy, technology, ethics, and education in meaningful ways, we need to think of data in all its dimensions, and take thoughtful approaches to balancing the relative opportunities, costs, rights, obligations, and interests that data use raises. It's not realistic to follow the EU model and treat data privacy in a stovepipe that's entirely separate from the private and security functions of European governments. It's not ethical to behave like China and Russia and treat all data as a tool to be leveraged to maximize state control over the people. It's not sensible to follow the traditional American model of legislating data privacy in sector-specific or state-driven silos that ignore the ways in which data crosses industry categories and geographical boundaries. We can't think solely about being left alone, or insist on doing away altogether with the biometrics that make it more difficult to move through the world anonymously. We can't focus solely on educating youth, when research shows that older people are less digitally literate and more susceptible to disinformation and online hoaxes than their children and grandchildren. If we only consider traditional privacy impacts, we risk losing sight of all the ways that personal data is used to make decisions about us, to influence our thinking, to shape our behavior, and to impact our future. If we're going to think seriously about cyber privacy, we need to approach data as a multidimensional problem: the data itself is often multifaceted; the possible uses are nearly always complex; the balancing of individual rights and public interests is nearly always susceptible to a sliding-scale view; and, to be sensible about data, we need to consider all of those dimensions each and every time we're faced with a decision-point about digital information and what, if anything, we should do to restrict it.

DATA RESTRICTIONS HAVE COSTS

Privacy protections always involve trade-offs, whether those are reduced access to targeted advertising, fewer tools for legitimate national security purposes and law enforcement investigations, less public access to information about a person's past, or a myriad of other trade-offs that we might face in any given circumstance.

In thinking through these tradeoffs, sometimes absurd scenarios can help to frame a principle, allowing us to start from an extreme outcome and work backwards to find a more sensible balance from there. The following hypothetical may be helpful as part of the framework for how to assess data privacy obligations. DNA can be extracted from feces; feces pass through sewage systems and waste treatment plants. We don't object to the Environmental Protection Agency testing water quality from sewage systems and waste treatment plants, or testing water quality in streams where storm drains or other systems have overflowed, pumping raw sewage (including feces) into the water. It's far-fetched to think that the EPA would extract feces from such water and then take the further step of testing it for DNA, which could point to everything from the identity of the pooper to a broad range of medical conditions and inherited tendencies, and even be used to trace the pooper's lineage. As a matter of public policy, we recognize that—although theoretically possible—that chain of events is unlikely: among other reasons, the EPA lacks the resources to carry out that kind of intrusive testing. (When privacy results from resource constraints, it's often referred to as "structural privacy.") And since there's societal benefit from having the EPA make sure that the water supply is clean, we're willing to accept this attenuated-but-possible risk of the privacy intrusion that could ensue. We could make other policy choices: we could prohibit EPA testing of water quality in order to avoid the hypothetical risk that the EPA would carry out privacy-intrusive DNA testing, but such a law or regulation probably wouldn't meaningfully advance privacy, and it would come at a high cost other social values, since we would lose the public health and environmental benefits that come from water quality testing.

It's useful to apply this kind of thought experiment to other kinds of data privacy challenges as well. If we took an extreme formulation, what privacy harm could result? What are the countervailing interests? If we took the approach that most fully protects privacy, what would be the likely harm to that countervailing interest?

To take another example, when it comes to national security costs, two kinds of security are frequently at odds: the security of specific pieces of information that many individuals would not want to see come into government hands; and the security of the nation, which requires that government be able to use at least some kinds of commonly available, as well as specialized, information in order to do its primary job of keeping the nation and its people safe. In considering the cost to individual privacy of any particular form of data collection or use, we should also be actively considering what hidden, shared costs may be associated with limiting that collection and use. In any given instance, the societal costs can include a reduction in law enforcement effectiveness and public safety as police departments lack access to information that could protect the public or help solve crimes. It might include the innovation costs associated with restricting the development of new technologies and business models that could provide welcome goods and services to the public. And it could include the dollar costs associated with implementing effective regulatory compliance regimes, a set of costs that are being keenly felt around the world right now as companies attempt to comply with the obligations of the GDPR.

DATA RESTRICTIONS MUST BE IMPLEMENTABLE

Privacy protections will always impede operations to some degree, and data security mechanisms will always bring with them some measure of inconvenience. Practical experience shows that when compliance costs are disproportionately high, regulated entities are more likely to resist. In addition, when the required compliance measures are particularly complex, it's more likely that even well-intended organizations will fall short, as more complex compliance regimes are frequently more error-prone. It's important to keep these practical realities in mind when policymakers are establishing privacy and security obligations; when regulators and overseers are examining the effectiveness and error rate of any particular compliance programs; and when privacy and compliance professionals are designing or advising on privacy and security controls and mechanisms. In privacy regulation and compliance, as with other things, the perfect can sometimes be the enemy of the good. As a result, in most situations and for most kinds of data, it helps to be practical, and to recognize that in many situations a modest protection that can be practically followed and readily enforced may be more useful overall than a more stringent protection that is extraordinarily difficult to comply with and therefore creates incentives to ignore or work around.

MEASURING DATA RESTRICTIONS BASED ON WHETHER THEY ENHANCE PRIVACY IN MEANINGFUL WAYS

In some cases, the question isn't about trade-offs; it's about utility. There are instances in which data privacy regulations impose societal costs without demonstrating a meaningful increase in privacy. For example, Poland's privacy regulator (a government agency) levied a 220,000 euro fine on a data broker for collecting publicly available information without providing individualized notice to each person whose information it obtained. Ironically, it appears that much of the data was gathered from websites of other Polish government agencies, such as a registry of business filing and court opinions. According to the Polish regulator, the data broker was obligated, under GDPR, to send hardcopy letters via the postal service to each individual notifying them of the intended use and their right to object, and to request their consent prior to downloading the information that was already publicly posted about them elsewhere. This hardcopy notice-and-consent process was required, according to the data protection authority, even though the cost of mailing would likely have run into millions of euros. The regulator's fine raised eyebrows across the

European data protection community as the first major fine on a data broker.[1] The Polish regulator had acted within the letter of the GDPR, but it isn't clear that there will be any actual privacy benefit to the individuals involved, since the information gathered by the data broker was already publicly available, and will continue to be publicly available after the fine.

In another example, European data privacy laws require websites to provide information about their use of cookies—these are the pop-up notices that have become nearly ubiquitous across the internet. The typical notice tells visitors that the website uses cookies; provides some general, high-level language about the fact that the cookies may track the user's behavior on that website; and may also notify the user that the site uses third-party cookies that could track other information about the sites they visit before and after this one. Most notices also advise that if a visitor continues to use the website, they will be deemed to have consented to the cookie data collection, and that if they don't want the site's cookies to track their behavior, they should stop using the website. (A smaller number of sites allow visitors to decline cookies but remain on the site, usually with a warning that the site's functionality will be reduced.) Given the generic nature of the explanations about what information is tracked, and given the take-it-or-leave-it choice that users face, it isn't clear what function the cookie notices truly fulfill, or whether they improve user privacy in any meaningful way. Meanwhile, companies expend resources to implement the pop-ups and make sure they comply with the latest laws; users are annoyed by having to click "I accept" or "x" to get the pop-up dialogue box to disappear. A great deal of energy and attention has gone into drafting and implementing cookie notice laws. But it's an open question whether anyone's privacy has actually been increased.

A third example provides a multi-jurisdictional perspective: "broadcast" social media platforms like Twitter and Instagram where most users intend their posts to be unrestricted and publicly available. The posts include a wide range of "personal data," including photographs, location information, contacts, and acquaintances—but, unlike posts on Facebook or other platforms that might be restricted just to "friends"—users on sites like Twitter and

Instagram often want their posts to have the widest possible reach. These posts sometimes include information that would be considered sensitive data, subject to additional protections under European law—political views, religious affiliations, sexual orientation, and the like. Under US law, because the user posted it publicly, the content is largely unencumbered by privacy laws. Under EU law, however, the collection and processing of that information may be heavily regulated, even though it was voluntarily and knowingly cast out into the world.

It's worth noting that in Russia, China, and other repressive regimes, government agencies frequently order platforms to remove offending posts and, sometimes, to deactivate the associated accounts. These heavy-handed restrictions are useful examples of the ways that governments can misuse publicly available information. But when a person has chosen to widely and publicly distribute information, the European approach of preventing others from collecting it seems to have little meaning. A better way to prevent its misuse would be to focus on prohibiting persecution or oppression—whether by government or private actors—based on those public posts. In other words, the protections should be on the back end (how the data can be *used*) rather than restricting collection at the front end. In that approach, collecting the data wouldn't intrude on information that the person wanted to keep confidential or limit access to, but persecution based on those public posts would violate privacy by using personal data to interfere with an individual's autonomy of thought, expression, and belief.

With those thoughts in mind, what kinds of privacy protections could we look toward?

MULTI-VARIABLE ANALYSIS AND INTERSECTIONALITY

Many privacy-protection remedies are still being proposed within stovepipes. People who are concerned with national security and law enforcement look at the problems through the lens of the Fourth Amendment. Data regulators in the United States look at them as a consumer protection issue, or through the sector-specific lens of existing laws. State legislators continue to expand the categories of information that trigger an obligation to notify state residents about a breach, but in most cases those laws provide individuals with few avenues for monetary relief and seldom acknowledge the importance to privacy of noneconomic data like personal photos, preferences, contacts, and affiliations. Class-action plaintiff lawyers continue trying to persuade courts that common law invasion of privacy should be legally compensable even when there's no resulting identity theft or fraud. Defense lawyers argue that even breach of statutorily protected data like Social Security numbers and payment card information shouldn't give rise to monetary damages because most Americans have now fallen victim to so many data breaches that there's no way to be sure that any subsequent harm stemmed from a *particular* data breach. Meanwhile, European data privacy regulators are focused on use limitations, notice, and consent as primary protections for personal data, but those restrictions have no applicability to the police or national security apparatus, creating a stark divide between the obligations imposed on private-sector use of data and government use of data for other purposes.

In some respects, the landscape resulting from these stovepipes (or "cylinders of excellence," if one wants to be tongue-in-cheek) looks grim. State-regulated data breach legislation has little or no intersection with criminal law. National security principles have little or no intersection with privacy law in specific economic sectors, like health care or banking. Platform providers defend their behavior with the absolute protections of the Communications Decency Act with no obligation to consider negative impact on a victim or potential societal costs. The Intelligence Community has learned hard lessons in intelligence oversight, but those haven't spread to state and local law enforcement, border regulation, or the private sector. Traditional banking information is regulated, but information volunteered to micro-lenders who rely on social credit scores from an applicant's personal contacts and social media interactions largely exists outside the bounds of banking laws. Schools and employers have legal obligations to maintain the confidentiality of certain kinds of data, but they are implementing increasingly intrusive surveillance on students and workers in ways ranging from cameras that measure engagement and attention to algorithms that make job recommendations and predict the likelihood of academic success. Social media platforms are creating personality profiles, our devices are feeding them data, and CEOs across Silicon Valley are rolling out products and websites that are exploiting our time, our attention, and our labor all for the purpose of carrying out unregulated social science research that will never be shared with the world but whose findings can be used to refine their product offerings and increase their future earnings.

The landscape of data-driven technologies is a complicated one, and nearly every area of our lives is touched by this intersection between ourselves, our rights, societal interests, and our data. Each of the current stovepipes—the different contexts, the different legal regimes, the different technologies, and the different uses—has lessons for the others. If we're going to make meaningful progress in appropriately protecting the rights of individuals while preserving the legitimate interests of society, all while technology continues to evolve, we need to shift our thinking. These technologies and the uses for our data operate in interconnected ways. In order to tackle these challenges, our mindset needs to be intersectional as well.

REORIENTING OUR FOCUS: HOW MIGHT WE APPROACH NEW LAWS?

The process of shifting the way we think will require creativity, open-mindedness, and flexibility in trying a variety of approaches to determine which ones suit which situations best. Assuming that we succeed in the micro-level tasks noted above, we will have created new, more nuanced, more situationally specific definitions of privacy that are rooted in digital reality. We'll be doing a more realistic job of asking whether particular privacy regulations in fact achieve their stated goal of meaningfully enhancing personal privacy, and we'll do a more concrete, realistic, and tangible job of recognizing that data restrictions have costs—and that we shouldn't impulsively impose privacy regulations any more than we should reflexively shy away from them. With these frameworks in mind, we can move toward taking a macro view, and thinking about what a universe of new laws might look like.

DO WE NEED NEW LAWS LIMITING DATA COLLECTION?

Individual actions and behavioral economic research both show that people often don't mind having their data used in new ways, so long as they benefit from it in some form. Laws that allow people to object to the collection and processing of *their* data may in some cases impose burdens on companies, but they allow individuals to continue enjoying the benefits of any kinds of processing that they personally find beneficial—and that's as it should be. The laissez-faire attitude that many people take toward their data provides ample motivation for companies to continue using broadly worded, open-ended statements in their privacy notices about what data will be collected, and for what purpose, and how it will be used in the future. The data market will continue to create incentives to develop new technologies that carry out front-end collection for wide-ranging purposes; if the reasons for the initial collection are broad, then wide-ranging follow-on uses can be allowed. There's nothing inherently wrong with this—innovation brings new products, new services, and new possibilities to individuals as well as to organizations and societies, and given that many uses of data are benign or openly welcomed by individuals, there's no good reason for government to erect artificial barriers to those uses by requiring, for example, that companies establish complicated regimes for asking individuals' permission to use their data in new ways, if the new purpose is just as benign as the original purpose was.

Equally important from a pragmatic viewpoint, it may be too late to close the barn door on data collection. A number of thoughtful commentators have suggested that, in today's digital environment, this may be a lost cause.

Instead of focusing on the point of collection, the trick may be in reaching consensus about how we define what uses are benign. For example, maybe the law should presume that any use that differentiates among people based on attributes like gender, ethnicity, or sexual orientation or identity should be presumptively illegal. The difficult question, though, is how to implement a standard like that, knowing that, in some cases, this differentiation will make sense. Processing information about a person's gender may be essential to determine which health-related products, services, and information are most likely to be of interest to them. Women might, on average, be more interested than men in information about ovarian cancer and menopause; men might, on average, be more interested than women in information about testicular cancer and male pattern baldness. But if a person's gender is also used to perpetuate the marketing of other kinds of information or products or services to them—which job advertisements to display in their search results, or what loan rates to offer—then is there a point at which the particular use of the data crosses a line? Undoubtedly, yes.

The challenge will be the difficulty of articulating the spot where that line has been crossed. It isn't at the point of collecting a person's gender or gender identity. It isn't at the point of the data tagging that makes it possible to populate a field in a broader set of records, to make clear in a data object or in a database what information is known about a person's gender. It isn't even necessarily at the point of running an algorithm against the gender-related data; since we can already imagine permissible uses of gender-related information, a sensible set of rules shouldn't prohibit running any and all analytics that pull on that field of information. So, at what point in data collection and processing do—or should—questions of law, policy, and ethics arise?

REDEFINE WHEN PRIVATE DATA USE BECOMES A LEGALLY REDRESSABLE PRIVACY WRONG

Until recently, courts in the United States have been reluctant to recognize a right to sue for unauthorized access to or exposure of personal data unless the individuals affected can point to actual damages. This reluctance stems from a US Supreme Court case called *Spokeo v. Robbins*, which holds that plaintiffs must have suffered an "injury in fact," which the court defined as invasion of a legally protected interest that is "concrete and particularized" and "actual or imminent."[2] When it comes to data privacy cases, most federal courts in the country have interpreted that requirement to mean that plaintiffs have only suffered a compensable injury—meaning they can only sue for damages—if they can show that the data breach involved something like credit cards or Social Security numbers that could impact their credit score or result in fraudulent charges or identity theft, and even then, some courts have been receptive to arguments from corporations that there have now been so many data breaches that plaintiffs should only recover for damages if they can show that they suffered an injury from this *specific* breach.

One of the consequences of this approach is that—for the data that many people consider to be far more personal than credit card or Social Security numbers—courts have been reluctant to recognize any remedy. One of the reasons why the Facebook–Cambridge Analytica program described in an earlier chapter garnered so much news and prompted such backlash against the social media giant is because many people view their private social media posts as more personal and more sensitive than the account numbers that courts have been willing to recognize as protected.

Within the United States, the courts in the Ninth Circuit have been most willing to recognize the possibility of a compensable privacy injury from data uses that don't involve a "traditional" data breach. In a recent opinion in *Patel v. Facebook*, the Ninth Circuit Court of Appeals ruled that, because Illinois had passed a law giving its residents the right to opt out of having their biometric data collected, residents had a right to sue Facebook for having deployed its facial recognition algorithms to "tag" users without first obtaining its users' consent.[3] It was a groundbreaking decision that led Facebook to settle the case for $550 million. (These protections are only available to Illinois residents.) In a separate class action lawsuit against Facebook for sharing user data with Cambridge Analytica, the district court, also in the Ninth Circuit, held that "when you share sensitive information with a limited audience (especially when you've made clear that you intend your audience to be limited), you retain your privacy rights and can sue someone for violating them." Further, the court wrote, "A privacy invasion is itself the kind of injury that can be redressed in federal court, even if the invasion does not lead to secondary economic injury like identity theft."[4] The Ninth Circuit's view in both these cases is far more privacy-protective than the bulk of courts that have required plaintiffs to show economic injury in order for any data-related violation of privacy rights to be eligible for monetary compensation. Yet, from a commonsense perspective, many people would find the Ninth Circuit approach to be more consistent with their own views about the kinds of information they most care about. As our digital data becomes more comprehensive and complex, this emphasis on economic injury appears to be increasingly out of sync and out of date with people's lived experiences.

In response to this disconnect, some legal scholars have suggested that it's time to reinvigorate the Victorian-era tort of intrusion on seclusion and update it for the digital age. Jane Bambauer, in an article titled "The New Intrusion," argues that there are four stages of interaction with data that are legally relevant, and where the law could regulate data use or provide remedies when it's misused. These are observation, capture (when a record is created), dissemination, and use.[5] Courts already recognize a right to sue when, for instance, a photographer takes photos of nude sunbathers who are on private property and then publishes them in a newspaper or online. Clearly, the individuals were visible; otherwise the photos couldn't have been captured. But they were in a location that they expected to be private, and with some exceptions, there's little legal justification—no societal value—in allowing the photographer to publish the revealing photos unless the subjects agree to it. A claim for digital intrusion on seclusion would recognize similar harms: information could only be captured, used, and shared without permission under circumstances that society as a whole considers decent and reasonable. This approach would allow individuals to say, in effect, that "just because I shared this information with my friends or uploaded my diary into the cloud storage associated with my personal webmail account doesn't mean I gave permission for it to be shared with the entire world—or, indeed, with anyone else." This position is consistent with the Ninth Circuit ruling in the Facebook–Cambridge Analytica litigation, and aligns with what, for many people, seems like common sense.

In some respects, this discussion brings us full circle back to the *De May v. Roberts* case, when the doctor brought his friend into the Roberts' home where the friend was privy to intimate sights, sounds, and moments. Although the court didn't use the words "invasion of privacy" or "intrusion on seclusion," the judges were clear that there are limits on what decent society should tolerate in the way of intrusions by strangers into what is private and what strikes deeply at our sense of ourselves. Reviving intrusion on seclusion for a digital age could provide one important remedy for individuals whose information has been captured, used, or shared in ways that they didn't expect or agree to, and in ways that other reasonable people would agree are wrong.

ARTICULATE FOURTH AMENDMENT FACTORS FOR GOVERNMENT SURVEILLANCE

A new and clearer framework could be established for government surveillance to comply with the Fourth Amendment. A number of the necessary factors have been alluded to in various cases over time, but—as the Supreme Court's 2018 decision in *Carpenter v. United States* makes clear—they haven't been drawn together into a coherent and unified framework. Under a proposal suggested by legal scholar Rachel Levinson-Waldman, factors to consider in evaluating whether a particular public surveillance activity is permitted by the Fourth Amendment should include:

1. The duration of the surveillance
2. The extent to which the surveillance uses technology that gives police an "unfair" advantage by greatly lowering the cost of what would otherwise be resource-constrained human surveillance
3. The extent to which it involves recording the physical location and movements of a person or group
4. The extent to which it involves gathering information from places that have an expectation of privacy, such as a person's home
5. Whether the technology in some other way undermines core constitutional rights (for example, current facial recognition technology that has discriminatory impacts on persons of color)
6. Whether multiple surveillance technologies are "piggy-backed" on each other

By explicitly analyzing each of these factors, Levinson-Waldman argues that government agencies may be better able to make sound policy decisions about which technologies to use, and under what circumstances, and how. Courts may be able to provide more consistency and certainty about what is permissible. And individuals may have some of their fears of overly intrusive surveillance assuaged.[6]

Regardless of whether this formulation is the perfect one, work like this makes clear that the time is long past for us to consider similar multi-variable analyses in assessing the privacy impact and potential legal restrictions and remedies relating to the intersection of privacy and technology.

ALLOW PARTS-PER-MILLION OF PRIVACY INTRUSION IN APPROPRIATE CONTEXTS

In organizations that have privacy-protective data-handling schemes, many—perhaps most—privacy violations result from human or technical error, not malicious intent. Yet many privacy restrictions take an all-or-nothing approach: a violation of privacy law is categorical, viewed in binary terms, rather than measured on a sliding scale of culpability and consequence. This is particularly true in the context of heavily regulated schemes, like HIPAA, FISA, GDPR, and state data breach laws. While this zero-tolerance threshold is the most privacy-protective position for individuals, it also imposes considerable costs, as it's extraordinarily difficult to drive the error rate to zero for any kind of complex data handling.

In contrast, the US Environmental Protection Agency (EPA) regularly measures levels of pollutants and toxins in our water, air, and soil. As a matter of public policy, US legislators have taken the view that it isn't feasible to reduce toxic contaminants to zero. Although it might be theoretically possible, we have a shared, societal acknowledgment that it would be exorbitantly costly to achieve a nondetectable level of even the most dangerous pollutants. We've looked at our collective budget of environmental resources and concluded that, while zero-level toxic outputs might generate marginal improvements in environmental health and safety, the costs are societally unsustainable. Consequently, we accept a certain level of risk—the risk of code yellow air pollution days, for example, knowing that we'll experience occasional, unacceptable code red days as well. Over time, our understanding of these risks has become more sophisticated, such as today's more widespread awareness that environmental burdens fall disproportionately on different groups of people according to measures like race and socioeconomic status.

So, some levels of pollution are inevitable and even acceptable. The discussion isn't about whether to drive the risk tolerance to zero, but whether allowing 100 or 200 parts per billion of a particular pollutant in smokestack emissions demonstrates a more appropriate balancing of the relative risks and costs that society is deciding how to trade off.

We make similar public policy judgments when it comes to airport security. The Transportation Security Administration (TSA) procedures can seem grindingly slow when travelers are pressed for time. But as a matter of public policy, we've decided we'd rather suffer the personal intrusion and inconvenience of having to take off our shoes and stand still for full-body scans than suffer a heightened risk of an in-flight catastrophe wrought by a bombing or other violent incident on a plane. Having reached that broad policy consensus, we then make further balancing refinements within our overall risk-balancing approach. Children under twelve don't have to remove their shoes, nor do elderly people above a certain age. Enrollees in the TSA PreCheck program may—with a background check and for a fee—receive expedited processing and lesser scrutiny, virtually ensuring a quicker trip through the security line. Private vendors like Clear have seen the business opportunity in this risk-balancing, offering, for a fee, a program of fingerprinting and identification that offers semi-expedited service—faster than the regular lines, even if it can't provide the reduced screening of the PreCheck lane.

Computer security experts have long known this balancing to be an integral part of their job in delivering effective network defense. The most secure computer, after all, is the one that's never connected to the internet, that has no wireless or Bluetooth capabilities, that has no drives for external media, and no CD or DVD drive. This computer has nearly perfect security. But that computer is also of very limited use. Similarly, computer security experts know that, in the event of a breach, it's seldom the case that deactivating an entire network is the right answer. These actions would stop the intrusion in its tracks, but they would also stop all manner of other vitally important processes. Not all intrusions are catastrophic, and for those that are relatively benign, the network security team knows that business needs likely dictate that the intrusion be addressed as expeditiously as possible while keeping available systems running so that business units can continue to carry out their mission.

In other words, there are a whole host of arenas, including environmental protection, travel safety, cybersecurity, and more, in which we balance competing interests—such as cost, convenience, impact on individuals, government interests, business operations, usability of networks and devices, feasibility and difficulty of technical security measures—in deciding exactly what societal risks we're willing to take. Often, the rationale for imposing ironclad, binary-standard protections and all-or-nothing privacy rules is that to the one person whose data is treated differently as a result of an error, that error can make all of difference in their sense of seclusion, dignity, and autonomy. But it's worth asking why we are so risk-tolerant in other areas, and so risk-averse in this one.

Right now, when it comes to certain kinds of privacy-related failures, we don't make many concessions to the inherent difficulty of maintaining perfect privacy of all personal data. Given the ever-increasing scope and complexity of the personal data and privacy landscape, we should be willing to consider making allowances for some kinds of errors that occur in certain kinds of privacy-impacting activities, which would, in an environmental context, amount to parts per billion.

Why should we allow some sort of acceptable margin of error? Because in some cases, we are so concerned about the hypothetical risk of malicious intent or harmful misuse that we overcorrect on the side of privacy, driving up compliance costs and causing otherwise reasonable efforts to grind to a near halt without actually increasing substantive privacy protections.

SOME GENERAL PRINCIPLES

Although it's possible to make specific suggestions on existing and proposed laws, it's a near certainty that, as technology evolves, we'll need to consider a wider range of possible new laws. Close behind that assumption is a second near-certainty: that it's unlikely that any of us can anticipate what all of those future needs will be.

With that in mind, the following section offers a handful of general principles that should serve as steady guides, or at least be considered in the equation, as we look toward continuing to test new hypotheses and develop more rigorous theories—along with laws, policies, and philosophies—of privacy over time.

WHAT HAPPENS ON THE INTERNET RESONATES IN REAL LIFE

Despite the proclamations of early internet enthusiasts, the ways that our digital data gets used has tangible impacts on everyday life, both on- and offline. Where the "Declaration of the Independence of the Internet" sought to declare the internet as a world of pure thought, independent of physical form, the experience of the past quarter-century has demonstrated, time and again, that what starts as online activity involving personal data almost inevitably has impact on our personal lives. Doxing, swatting, and revenge porn result in real-life humiliation, reputational damage, and mental and emotional distress for the victims. Online political disinformation manipulates public opinion in ways that can lead to real-world demonstrations, a deepening of offline societal divisions, and even the swaying of the outcome of elections.

Some commentators will argue that these kinds of destructive uses of personal data are simply reflections of human nature—that similarly malicious actions can, and did, take place long before the advent of digital media, and the proliferation of personal data made it possible to target specific individuals for harassment, manipulation, or other forms of messaging. Those commentators would be correct, to a point: malicious actions and destructive motives have a long history and may be inherent in human nature. But we don't need to make it easy for modern technology to accelerate or amplify those worst tendencies. We may not be able to change human nature, but we can impose sensible limitations on people's ability to exploit the power of data-driven technology for purposes that are deliberately harmful to individuals or that undermine important interests of society. With that in mind, the following principles may be useful guides.

STRIKE A THOUGHTFUL BALANCE ON THE FIRST AMENDMENT, AS WELL AS THE FOURTH

When it comes to protecting individuals' Fourth Amendment rights, there is an extensive body of case law, regulation, investigations, reports, and scholarship on foundational principles, practical application, and oversight and accountability. Not everyone agrees that the balance has been correctly struck, but there's a clear framework within which laws and policies can be updated and changed in order to strike a socially constructive and legally principled balance between government surveillance and individual rights.

When it comes to balancing speech against privacy, however, Sec. 230 of the Communications Decency Act has stood as an impenetrable shield against any attempts to impose some degree of accountability on internet platforms and websites that publish defamatory, harmful, or intrusive content. The effect has been the cultivation of a thriving ecosystem of revenge porn, defamatory content, political disinformation, and deep fakes. The United States' bedrock commitment to protecting speech will continue to set it apart from authoritarian governments around the world. But the balance that has been struck with Sec. 230 leans so far forward that it not only protects private platforms that profit from knowing publication of false, defamatory, and harmful speech; it actively encourages it through making clear that online platforms can publish—and profit from—content that would be legally actionable if it were published in other, offline mediums. The results are frequently devastating to individuals, as the law elevates freedom of speech disproportionately far above any other concerns, including the privacy of the victims whom revenge porn, disinformation, and other forms of harmful speech are targeting.

FOCUS ON DATA TYPE, NOT DATA SOURCE

Under many pillars of US law, the protections applied to data don't derive from the inherent sensitivity of the information itself, but from the manner in which it was acquired. So, for example, health-related information is protected under US privacy laws if it was collected by a doctor or hospital, but not if it was collected by a fitness wearable or an app. Similarly, location information may be protected from government use if it's collected from a mobile phone, but not if it comes from video surveillance cameras. There are often differences according to the nature of the data collector or type of use, with government uses generally being much more highly restricted than in the private sector, but these distinctions don't always make sense.

In contrast, the EU's privacy law framework tries to correct this by protecting all types of personal data, but it suffers from other shortfalls, as it doesn't protect personal data for all purposes, and in some cases, it restricts data in ways that seem illogical. So, for example, the EU's privacy laws impose stringent restrictions on organizations and individuals, but exempt law enforcement or national security activities from those regulations. The very uses that have raised the most concern in the United States—and that have caused the EU to harshly criticize the US—aren't addressed by EU law with respect to their own member states. If privacy is truly a fundamental human right, then the fact that European Union law offers no protections against overreach by its own intelligence and law enforcement agencies seems like a significant oversight. In other instances, the EU penalizes activities, like collecting publicly available information from government websites, that seem to create low privacy risks. And overall, there are a number of respects in which EU law tends to impose high regulatory compliance costs for relatively low privacy benefit, as in the case of cookie notice requirements. In other words, the EU's approach may intend to protect all types of personal data through a broad definition, but its approach also results in anomalies that don't seem to advance privacy or other social goods.

Whether the information is about our movement and location over time, health conditions, web searches, personality traits, income, educational achievement, work status, relationships, interests, or any of the myriad other aspects that go into personal data profiles, privacy regulations shouldn't narrowly focus on how the data was collected, but on what the privacy impact to an individual of that information might be.

FOCUS ON RESTRICTING USES, NOT SOLELY ON RESTRICTING COLLECTION

In separate conversations, two leading thinkers on privacy in the digital world came to the same conclusion: meaningful privacy protections are more likely to result from regulating how data may be *used*, rather than on how it is acquired. One of the areas in which this is most evident is in the context of targeted data gathering that results in some amount of incidental collection—a scenario that's most often relevant to government activities that are regulated by the Fourth Amendment in the United States, or by various statutes and state constitutional provisions around the United States and overseas. Although many private-sector activities are only loosely regulated today, the stringent oversight and compliance mechanisms that are in place for some kinds of government data collection can provide lessons on ways that controls on data processing and use can counteract the potential intrusion that might otherwise result from the mere fact of collection itself.

Since 1977, when FISA was passed, lawmakers have recognized that collecting relevant information will sometimes also result in collecting data that has no intelligence value. To address this problem, Congress enacted a framework in which all three branches of government have responsibilities for creating, implementing, and overseeing stringent protections on data retention, processing, analysis, and dissemination. Although FISA is frequently criticized, its compliance framework is robust and effective in bringing errors to light.

One example is the debate over incidental collection under FAA Sec. 702, discussed earlier in the book. Congress recognized that in order to allow vital intelligence collection to happen, some incidental information would be acquired along with the intentional, targeted collection. One way to address the privacy issues associated with incidental collection would be to adopt the position that some privacy advocates argue for; however, requiring that incidentally acquired information be masked, sanitized, or destroyed imposes significant resource and privacy costs. In an undifferentiated data pool, particularly a set of unstructured content, it's costly to find that data, and even more expensive to mark, mask, or segregate it. From a business operations perspective, it drains time and energy to undertake these efforts, and to train people to identify the information so that it can be handled separately. It means directing analysts and technology developers away from the core work of identifying information that *is* relevant to their work, and redirecting their efforts toward identifying the information that isn't relevant at all. Exacerbating the challenges, it's privacy intrusive to hunt for the incidental information that is lost within a vast data set and that isn't otherwise likely to be scrutinized. Instead, the government has in place a robust framework of training, spot checks, auditing, and review that governs how incidentally acquired information must be protected when an analyst comes across it.

Although not all privacy advocates agree these measures are sufficient, they nonetheless go far beyond the programs in place in most private-sector organizations. The private sector will continue to expand its data collection, so privacy advocates can look at the ways that government oversight frameworks and minimization procedures are used to protect incidentally acquired information for constructive examples that could be adapted to nongovernmental contexts as well.

FOCUS ON PREVENTING OPPRESSION AND PERSECUTION, NOT SOLELY ON PREVENTING PROCESSING

The lessons of the most brutal regimes in history and around the world underscore the ways in which government can use personal data to oppress and persecute people within their borders, leading to internment and even to systematic genocide and pogroms. Despite the lessons of these atrocities, it's not clear that preventing governments from collecting or cataloguing certain kinds of personal data is a sufficient—or even effective—tool to prevent future atrocities from occurring. In one stark example, the Rwandan genocide was made possible in part by the fact that colonial history had deepened the sense of distinction between different ethnic groups. But the Rwandan government didn't need to maintain an ethnicity registry in order to carry out the genocide (although, to its shame, it did). The truly necessary underpinnings were for the government to foment a sense of otherness over a period of time, then create laws enabling or requiring the slaughter that later ensued. Ethnic designations on identity cards helped contribute to the sense of otherness. In many cases, however, the killings weren't carried out directly by the government, but by private citizens; and the extremist Hutus who slaughtered their Tutsi neighbors didn't need to consult a government database or registry, or purchase advertising lists from a data broker, in order to identify the Tutsi targets that attacked and killed.

In Nazi Germany, the persecution and slaughter were carried out in a systematic and government-managed fashion, with the security services creating and maintaining lists of people who fell within undesirable groups or who demonstrated disfavored characteristics. There's no question that the ability to amass and catalogue this information—whether obtained by police observation, tips from informant neighbors, or other sources—greatly facilitated the Nazis' ability to carry out its extermination plans. But while the information was key to the Nazis' implementation of their ethnic cleansing plan, the logical connection doesn't work the other way. In other words, all persecution can be fostered by collection and processing of sensitive personal data, but not all collection and processing of sensitive personal data leads to persecution. In fact, the opposite is true: in some instances, collecting information relating to race, ethnicity, religion, sexual orientation or identity, or disability can help support important, positive social goals, like recording instances of discrimination so that they can be remediated, measuring trends in discrimination to adjust policy responses, and assessing the effectiveness of interventions intended to increase overall equality in order to direct resources into expanding the efforts that work best.

Consequently, when it comes to protecting vitally important human rights and liberties that extend beyond privacy to core Western democratic ideals such as the right to life, liberty, and the pursuit of happiness, it will often be more important, and better advance society's overall values, to focus less on restricting collection and more on prohibiting and preventing abuse of power and persecution, whether at the hands of the government or the private sector.

In considering how to enhance existing prohibitions against, for example, actions that have racially discriminatory impact, the use of personal data could be incorporated into existing laws or added to new laws as an aggravating factor in any instances in which there is abuse of authority by government or private-sector potentates (such as employers and schools). It could also be an aggravating factor in considering damages in civil suits alleging direct violations of privacy-related rights, such as intrusion on seclusion, or in lawsuits for violation of other rights—such as the right to non-discriminatory treatment in healthcare, housing, and other private transactions—where misuse of personal data played a role. A number of other approaches could be adopted as well; what's essential is to consider carefully the possibility for beneficial as well as harmful uses of these kinds of data, and to fashion laws and policies that address uses, not merely collection.

FOCUS ON DATA DEMOGRAPHICS, NOT SOLELY ON INDIVIDUALS

The New Economics Foundation, a UK-based think tank, makes a persuasive case for considering a shift in the adtech ecosystem in its 2018 report, "Blocking the Data Stalkers."[7] Currently, when advertisers are looking to place ads through platforms like Facebook and Google, they frequently gain access to personally identifiable information of individuals. That data is then subject to leak, theft, and commodification. The NEF recommends changing the adtech process so that the bid requests issued by websites would send demographic characteristics rather than specific personally identifiable information to the advertising bid network. This might diminish some of the profit margin of advertisers and data brokers, but it would also increase privacy protections for individuals. Further, NEF suggests, it would "force tech giants to diversify their business model away from services based on constant surveillance and advertising" and "give power back to websites which spend time producing content and have a dedicated user base."[8]

The best solutions generally incorporate some element of compromise. Thus far, tech platforms have been reluctant to sacrifice their profit margin in order to increase individual privacy. Proposals like this one that attempt to balance those interests deserve further exploration to see whether and how they can promote a more satisfying equilibrium between privacy, commerce, and technology.

INCORPORATE ALGORITHMIC ETHICS IN DECISIONS ABOUT USE RESTRICTIONS

It's notoriously difficult to understand how machine learning algorithms reach their conclusions. What we do know is that many AI capabilities are trained on nonrepresentative data sets. This can lead to conclusions that aren't scientifically well founded or well grounded, such as in the case of medical studies that attempt to predict health outcomes for women based on data derived from studies that were performed only on men. It can also lead to the perpetuation of preexisting social biases, as in the case of algorithms that are intended to support hiring decisions or career paths, but that have the effect of reinforcing trends in employment and education in which there has historically been less recognition of the abilities of women, minorities, and other groups that have been subject to disadvantage and at times outright discrimination in the past.

A great deal of promising research and interdisciplinary investigation into algorithmic transparency and ethics is currently being done by groups like the AI Now Institute in New York. As algorithmic processing of personal data continues to expand, private-sector users, government agencies, and lawmakers would all be well advised to pay close attention to this ongoing work, to support its continuation, and to leverage its lessons learned in new technologies, programs, and regulations relating to personal data.

FOCUS ON THE USES THAT HAVE GREATEST IMPACT ON INDIVIDUALS' LIVES

Even when intended for benevolent purposes, personal data can easily be misused by employers and educators. One of the most pernicious facets of pedagogy is the reality that when children are told that teachers' expectations of them are low, the students perform down to those expectations. The same is true in the workplace: set high expectations, and people generally live up to them; give people a challenge, and they rise to the occasion.[9] Concurrently, when students are told that they're likely to fail, they often do, and when workers are treated in ways that undermine their dignity and autonomy, they seldom reach their full potential.

With this in mind, use restrictions should consider the ways in which data-driven predictions from schools and employers can impair individuals' ability to perform well or to reach their full potential. Policymakers and researchers should be equally alert to other kinds of power imbalance: to other types of situations in which information about a person might disadvantage them with respect to significant decisions or institutions, whether those are lenders and credit scores, eligibility for government benefits, probation hearings or eligibility for parole, and more. The list of potential situations in which personal data becomes part of a power play is likely to grow as data-driven technologies expand. Areas of structural imbalance should remain front and center in the ongoing assessment of what risks accrue to individuals, and what steps might mitigate them.

BE WARY OF POTENTIAL HARM, BUT LEGISLATE FROM FACTS RATHER THAN FEAR

When difficult balancing questions spring quickly into public consciousness, it can prompt a knee-jerk response, whether from advocates, legislators, or the public. The reaction to the NSA's Business Records FISA program, and the congressional debates that led to the passage of the USA Freedom Act, are a useful exemplar.

The BR FISA program is sometimes pointed to as an example of overly intrusive national security surveillance, a program that gets the balance wrong. According to declassified information, it was both much lower in proven intelligence value than its more focused counterparts like the FISA Amendments Act (FAA) section 702, and it had the potential to be much more intrusive.[10]

But was it really?

The BR FISA data consisted of communications externals only. There were concerns that it was theoretically possible for an NSA analyst to look at that data and marry it up with extensive external information in order to get a picture of a target's intimate details—for example, that a person's number contacted the numbers for a domestic violence abuse line, a religious institution, a pornography purveyor, or a particular medical provider.[11] However, there was never evidence that NSA was actually doing that. In fact, just the opposite constraint had been built into the program: before querying the data, analysts had to prove they had reasonable articulable suspicion that a particular number was being used in connection with international terrorism. The data (which consisted solely of information derived from the kinds of billing and other records maintained by telephone companies that the *Smith v. Maryland* court had said didn't require a warrant) was kept strictly segregated from other data sets that could provide the kinds of corollary information that privacy advocates were worried about. What NSA was looking for were numbers potentially associated with terrorist planning and the numbers they were in communication with.[12]

The BR FISA program, like other NSA surveillance programs, specifically acknowledged the importance of First Amendment rights and sought to build in those protections.[13] NSA's training program reminded the analysts who were permitted to work with the BR FISA metadata that the First Amendment protects the free exercise of religion, speech, the press, peaceable assembly, and the right to petition the government for redress of grievances. Under the program, each Foreign Intelligence Surveillance Court authorization lasted no more than sixty to ninety days, and each order listed the specific international terrorist groups that were valid subjects of queries using the BR FISA metadata.

Analysts were prohibited from using the BR metadata to investigate any group that wasn't listed on the order—a protection to make sure they didn't use the metadata for purposes that were broader than or different from what the court intended to authorize.[14] The information NSA received from participating telephone companies was only a small subset of information that the companies held in their business records. The "telephony metadata" gathered under the BR program was limited to specific fields of information: the originating and terminating telephone numbers, three kinds of device identifiers, the trunk identifiers (signaling information) for the call, any telephone calling card numbers, and the duration of the call. Notably, the BR FISA data didn't include the substance or content of any communication, nor did it include the name, address, or financial information of a subscriber or customer.[15] A number could only be used as a seed to conduct contact chaining in the BR FISA data if an analyst could point to a distinct, articulable fact supporting the connection between that number and a terrorist group listed on the court order.[16] The training warned analysts that they weren't permitted to rely on instinct or hunches rather than facts, and provided a list of the kinds of factual information they could consider, including published intelligence reports and other intelligence information.[17]

There were robust oversight and compliance mechanisms built into the program as well. The court orders limited the number of analysts who were permitted to query the data, and each of those analysts had to be specially trained. The NSA has a legal standard for "reasonable articulable suspicions"; each RAS justification had to be approved in advance. Each query was audited, each IP address used to query the data was logged, and the date and time of access and user login for each authorized analyst was logged and reviewed.[18] Each query was also reviewed after the fact by overseers from the Office of the Director of National Intelligence and the Department of Justice. Any sharing or dissemination of query results from the BR FISA metadata was strictly and tightly controlled, requiring express approval from a limited number of NSA officials authorized to grant such permission.[19] Detailed reports were submitted to the FISC about all of NSA's activities under the program.

The controls on the data were significant, and far greater than was widely reported at the time. The program was repeatedly authorized by multiple federal judges, each acting independently, and congressional committees were regularly informed about it. And data from the program was tightly controlled to reduce the scope of potential invasion of privacy and to limit the chance of misuse of the data. None of these safeguards prevented commentators from charging that the program was unlawful, and ultimately it was transitioned to a different framework with explicit statutory authorization under the USA FREEDOM Act.

Setting any criticisms aside, none of the facts about legal review, judicial authorization, congressional review, constraints, training, and oversight change the important policy question of whether a future Congress should authorize a program of this kind under American law. But they do point to the importance of making sure that policymakers and the public have an accurate set of facts to work from when trying to make difficult or complex decisions.

PRACTICAL TIPS AND NEWS
WE CAN ALL USE

As individuals become more aware of the ways that information about them is generated and used, many people find themselves asking what practical measures they can take to align their data profile more closely with their values. That is, for individuals who are comfortable with robust personal data processing, how can they gain the greatest possible benefit from that collection and use? And for individuals who wish to restrict the ability of governments, employers, educators, and organizations to access information about them, what kinds of practical measures will allow them to remain aware of current data trends? How can they make sensible decisions and know what actions they can take to participate as fully as possible in both the physical world and the digital economy, and achieve a level of personal data protection that aligns as closely as possible with their values and their goals?

EDUCATING PEOPLE ON THE RISKS OF DATA-DRIVEN MANIPULATION

It seems to be in vogue for people to talk about the importance of teaching critical thinking in schools, at all levels of education. That view isn't wrong.

But while teaching critical thinking in schools is necessary, it isn't sufficient to combat the multitude of ways in which individuals can find their opinions and actions manipulated by sophisticated actors—foreign governments, political advertisers, consumer marketing companies, or others. Research like the studies being done at Northeastern is providing important evidence indicating that the people most vulnerable to these kinds of manipulative misinformation and persuasion campaigns are the ones who are furthest from school: older generations who, broadly speaking, tend to have a lower level of digital literacy, are often more socially isolated, and who often rely on fewer sources of information to make up their minds about any particular topic.

Critical thinking needs to be taught in schools. And in nursing homes. And in workplaces and senior centers, and through televised and radio broadcast public service announcement campaigns. The topic could be addressed in popular media like books, television news, and the plots of fictional television shows. Hollywood may have a better opportunity to reach older Americans than educators; news broadcast channels may have a better opportunity to do so than print newspapers.

Many people, including former Supreme Court Justice Sandra Day O'Connor, have lamented the decreasing degree of civics literacy in the United States and the fact that so many Americans have so little understanding about the core principles enshrined in the Constitution, or about how the American system of government works.[20] Any programs designed to enhance the degree of civics literacy in America would do well to include education on our nation's principles and system of government—and also on digital literacy. The inability to differentiate factual information from misleading propaganda poses a threat to our democracy itself.

AVOIDING BEHAVIORAL MANIPULATION CAUSED BY PLATFORM AND APP DESIGN

In 2019, legislation was introduced in Congress that would prohibit platforms and apps from forcing habit-forming patterns on consumers. The Social Media Addiction Reduction Technology (SMART) Act points to a number of very real mechanisms that data-driven technologies use to hold our attention and—by keeping our eyeballs on the screen—gather ever-increasing amounts of data from us.[21] Although the bill is unlikely to gain enough support to pass in its current form, the list of practices that it would ban provides a useful shorthand of developer techniques that we can individually be wary of. These include autoplay, the feature that causes the next video to automatically queue up in Netflix and YouTube, in recognition of the fact that we're more likely to continue watching if we're passively receiving new videos than if we have to proactively click on the screen, or the remote, in order to get the next video to begin. (User tip: many platforms have a setting that allows users to turn autoplay off. By consciously choosing when to have the feature enabled or disabled, we can take more control of our data, and our time.)

Push notifications are another feature used by apps, platforms, and media outlets to grab our attention and bring it back to the screen. Studies have shown that these notifications—the pings, the breaking news alerts—trigger neural activity that, in the moment, feels satisfying (who doesn't like a surge of dopamine?) but have the overall effect of creating stress, interfering with our relationships and interrupting our train of thought, leaving us unsatisfied and craving more. The apps, platforms, and outlets that barrage us with these notifications also have settings that allow us to turn the notifications off. In order to preserve your sanity, reduce your screen time, and limit the amount of personal data you're generating and giving away, be judicious in choosing which apps you're going to allow to send you notifications, and limit that privilege to the ones you really care about and get the most sense of value from.

Finally, any kind of app that offers in-app purchases, posts notifications at random times, or encourages records of unbroken activity pulls us in, pushes us to give up more data (and sometimes money), and risks wasting our time. Whether it's keeping a Snapstreak on Snapchat (a tally of how many continuous days two users send photos to each other via the app), logging in for daily rewards in Candy Crush, or checking Instagram obsessively to see when it will update our feed next, all of these are developer-created motivations that we should be wary of. There's nothing wrong with using any of these apps. But we don't want the apps to be using us—we want to be the ones consciously and freely deciding when to use them.[22]

CONCLUSION
MAKING SURE THAT HUMAN BEINGS STILL PASS THE TURING TEST

Today's data-driven economy and signal-driven world have had the effects, some would suggest, of diminishing our ability to think critically, read deeply, consider carefully, and connect intimately. If our personalities can be summed up by seventy likes or a persona-based profile of a thousand data points, and if we are so easy to manipulate and cajole, to influence and persuade, that all a political or commercial advertiser needs is access to data files of our digital activity, then what have we lost in terms of spontaneity, creativity, and our capacity to develop original ideas and to surprise ourselves and our fellows? If we are so wary of the risks of government surveillance that we fear to whisper the truth even in what we hope to be the privacy of our homes, or to walk with our chosen companions down a city street, then what have we lost in our ability to achieve the fullness of our human potential?

The risk, it seems, is that while computers become more able to process information with nuance and complexity, to "think" more like humans, humans are simultaneously becoming more like computers in our thinking. We're tempted to become more passive in accepting data inputs, and to think less rigorously about the information that's presented to us. Our thinking becomes increasingly susceptible to the old computer-programming adage: garbage in, garbage out. If a questionnaire calling for complex thinking or resonant emotion were to be presented as Alan Turing's parlor game today, with a computer and a person answering the same questions, would the human still be able to pass the Turing test? Would the evaluator reviewing the results be able to tell which responder was human and which one was not? That, it seems, is the crux of our challenge in this data-driven age: how to maintain the essentials of our humanity—the independence and agency, the dignity and sense of self, which rest on the bedrock of individual privacy—in a world in which more and more information created from and about us is released.

If these are some of the many risks of data, is there any more optimistic view to be found? Is there any hope at the bottom of Pandora's box? It may help to revert back to our analogy between quantum physics and quantum policy. In the non-quantum realm, in the world of Newtonian physics, mathematics has provided us with equations that, for hundreds of years, have been remarkably effective in predicting the vast majority—approaching 100 percent—of all actions and interactions in the universe. So it smacks of hubris to assume that we, as complex but nonetheless Newtonian beings, should somehow be above predictability.

On the contrary, the vast sum of human experience suggests that most of us are highly predictable most of the time. Psychologists, psychiatrists, and social scientists have created academic disciplines out of understanding and predicting human behavior. Criminologists, intelligence analysts, and political pundits have attempted to provide security and stability to our nation, and to shape its future directions, based on little more than their ability to gather data and use it to predict the behavior of individuals and groups of humans. Teachers, coaches, and employers have used personality assessments, aptitude tests, performance observations, and motivational talks to try to analyze, predict, and shape human behavior. Even at home, within the family, parents, partners, and children try to analyze, predict, and influence each other's behavior on a near-daily basis.

Poets and artists have tried to plumb the depths of the human condition. People in countless disciplines have devoted their lives and careers to attempting to predict the behavior of the people around them whose opinions and actions matter most. Although we don't yet have comprehensive systems for approaching the ethics, policy, and privacy implications of the fact that we're allowing machines to assist us in an endeavor that we've undertaken—in one form or another—for thousands of years, this gap in law and philosophy doesn't mean that the endeavor itself is new, or that we should reject out of hand the value that increased data sets and computing capabilities can bring to the task.

In a chance conversation on a recent long-haul flight, I was talking with a computer scientist who said, "Computers and machines haven't made us less human. But we've made machines more human." He went on to argue that if we look at the span of human behavior, the problems we're concerned about now are enduring ones. "Think about how long propaganda has been around," he continued, pointing to Sun Tzu's classic line in *The Art of War*: "The very idea of knowing your enemy as being critical to success in warfare—isn't that what every battlefield commander tries to do? Isn't that what everyone tries to do?"

Our conversation was a far-reaching one, and before long he was sketching on the back of the airline napkin a representation of a simple neural network, and the mathematical equations that underlie it. He pointed out that neural networks are already better at predicting some things (e.g., baseball signs) than people. He was more than just resigned to the idea; he seemed genuinely comfortable with it. "Neural networks are built to approximate human brains," he pointed out, "and as they get better, they'll become more and more like human brains. Eventually they'll be better at predicting things than human brains."

We continued to chat over honey-and-ginger ice cream, a dessert item that made me wonder which airline nutritionist predicted, based on which data, that the large and diverse group of economy-class passengers would prefer this choice to the mango sorbet we'd been offered on the previous long-haul flight. (My small-and-unscientific survey of passengers in the nearby seats and surrounding rows seemed to indicate that the mango had been more popular.) We continued talking well past the time when the cabin lights were dimmed. Data, I've found, is becoming like the weather: it's something that almost anyone can talk about. Everyone has some level of interest, it seems, in apps and smartphones and online shopping recommendations, in fitness trackers and surveillance and backup storage clouds; we all want to have a better idea of who knows what about us, and how that information is being used.

As our conversation went on, my computer scientist friend pointed out that "We're comfortable with the ability of computers to predict some kinds of things. We're even happy about it." The example he gave was the humble calculator: when computers first came along, it didn't take long to realize that they were far better than most people at doing basic arithmetic. We're happy to have calculators relieve us of the tedium of doing complex mathematical functions. And no one felt like that was making us lose our humanity.

"But when it comes to predictions about human behavior," I said, trying to find the right words, "or about things that call into question our sense of autonomy or our personhood, we become very uncomfortable."

In response, my friend offered two competing views: "From the perspective of a data scientist, or a computer programmer, this is an exciting time. Computers have always been better at some things than us. So one way to look at this is as another way that computers are advancing. Just like we got to offload the burden of routine arithmetic onto calculators, now social scientists can offload the work of analyzing human behavior onto computers."

As he said this, I made a mental note to run this theory past a social scientist and get their perspective and response.

"From a security perspective," my companion continued, "there are some real concerns. Suppose you get a neural network that's so good at assessing human behavior that it can use your social media posts to guess what your password might be. Maybe the neural network figures out that people who make certain kinds of social media posts tend to choose certain kinds of passwords. That could really undermine data security." He frowned. "Of course, you'd have to solve this by creating neural networks that are designed to generate better passwords that humans can remember but that are harder to predict."

While he was thinking about password-development AI, I chimed in again. Based on all my conversations with people who are thinking deeply about data, from all of my review of data privacy laws and practices around the world, from everything I've learned in writing this book, I said, "I don't think the answer can realistically be to try to put the data back in Pandora's box. We're not going to stop new data from being created. We're not going to stop the growth in the numbers and kinds of new data. And we're not going to stop people—whether its companies or governments—from collecting, aggregating, and analyzing that data.

"So it seems like the real challenge lies in creating sufficient restrictions to rein in the human tendency to misuse information for purposes that we've collectively decided are unacceptable in society," I said. "Those unacceptable uses should include suppression of political dissent; discrimination in employment, housing, educational, and other opportunities for people of particular ethnic groups, or with particular medical conditions, or belonging to particular religious groups, or of particular sexual orientations or gender identities; or for viewpoint coercion of any kind; or other similar discriminatory purposes."

As my companion put in his headphones and started listening to music, I went back to thinking about *The Art of War*. One of the passages that's become most famous in English-language translations is this one:

> If you know the enemy and know yourself, you need not fear the result of a hundred battles. If you know yourself but not the enemy, for every victory gained you will also suffer a defeat. If you know neither the enemy nor yourself, you will succumb in every battle.

Yes, perhaps in some respects data is our enemy. But maybe that isn't the point. Understanding data won't help us overcome the risks it presents to our privacy, our individuality, or our autonomy. In order to mitigate those risks, we need to understand ourselves. We won't be able to develop laws, definitions of privacy, or codes of ethics quickly enough to keep pace with all the ways that data is going to continue to grow. However, if we focus deeply on understanding all the ways that our human prejudices, lust for power, desire for wealth, and other frailties make us inclined to try to gain leverage over our fellows through the things we know about them, then we find ourselves on a path that leads to both big data and vibrant personhood—a path to quantum policy.

When I set out to write this book, I'd hoped it would be possible to articulate a sort of Grand Unified Theory of data privacy. A set of overarching principles that could guide us—morally, philosophically, ethically, legally—through the quagmire of questions raised by the ways our actions, interests, and identities are captured by the technologies that surround us—the same technologies we sometimes embrace, at other times loathe, and that frequently we feel we can't live without. It turns out that developing that kind of overarching, holistic view on data privacy is harder than it seems. The Grand Unified Theory may be a task better suited for a multivolume set, and this book provides a mere starting place. But my hope is that, in providing a catalogue of risks and relevant questions, along with a useful framework for thinking about the future, this book may spark further, future discussions—a down payment, if you will, on the multivolume set.

ACKNOWLEDGMENTS

The inspiration for this book came unexpectedly, during a class I was teaching to a group of Intelligence Community officers in 2015. I had provided similar lectures frequently over the years: an explanation of how electronic surveillance law had evolved over the years, how the Fourth Amendment intersected with the authorities framework for the IC, and how the evolving privacy landscape in Europe and elsewhere in the world took an approach toward private-sector privacy that was so different from the laws governing the US IC. Following the class, an analyst from one of the IC agencies approached me to ask for clarification. After we'd talked through his questions, he said, "You know, you really should write a book about this."

Over the next several years, I counseled countless clients as they grappled with the challenges of emerging data privacy laws and evolving privacy norms in the United States and around the world. I'd talked to companies, nonprofits, and government departments and agencies; I'd seen firsthand the disconnect between different standards across different industries, different data types, and different places around the world; I'd written dozens of articles and given countless presentations on data privacy and cybersecurity. In the course of the teaching and reading and practical experience, I was reminded on a daily basis how fast data-driven technologies were changing, how slow the law was to keep up, and how challenging it was for anyone who didn't live and breathe these issues as a full-time job to keep up with it all. In the midst of it all, a detour to serve as senior minority counsel in the Senate Intelligence Committee sharpened my focus on the role that oversight could play in these issues.

Through those experiences and observations, the shape of this book began to take form.

As with any large project, I'm indebted to more people than I can name here. Nonetheless, a few names in particular stand out.

My indefatigable agent, Laura Strachan, met me for coffee by Annapolis City Dock and was willing to take this project on when the outline was broader, more sprawling, and less focused than the book that ultimately resulted. Glenn Yeffeth of BenBella Books saw in that overbroad proposal from a first-time would-be author the kernel of a viable book. My editor, Vy Tran, pushed me along when my writing fell behind schedule and asked the questions that helped me see what work was needed in order to bring a rambling manuscript into focus. The entire production team at BenBella supported, encouraged, and facilitated this work with their professionalism and attention to detail, resulting in a much better final product.

The content was informed and richly shaped by the many people who were willing to let me do formal interviews or have informal chats with them about their perspectives on cybersecurity and data privacy. The University of Maryland students who took my classes on Information Privacy and Internet Law asked me hard questions that helped sharpen my thinking. The lawyers, engineers, computer scientists, analysts, and others who I've had the occasion to co-present with, co-write with, work alongside, and debate have all helped shape the thinking that went into this book.

Most of all, I owe an enduring and unrepayable debt to my family: to my husband, our children, and their grandparents. Dave, Summer, and James put up with what seemed at times like endless hours of me sitting huddled over a keyboard, surrounded by piles of research scattered on the floor. They pulled me away from the computer when I needed a break, and they prodded me back to it when I needed to return. They served as steady sounding boards for the ideas in this book, providing inspiration for key chapters and insights for the conclusions. They assured me that the work would get done, because it always does. Without them, this book wouldn't have been written. Words can't convey my gratitude to, and for, each of them. Without them, my heart would be immeasurably emptier, and my life would be incomplete.

NOTES

INTRODUCTION

1   Simon O'Dea, "Smartphone Users in the United States 2010-2024," Statista, April 8, 2020, https://www.statista.com/statistics/201182/forecast-of-smartphone-users-in-the-us/

2   "Mobile Fact Sheet," Pew Research Center, June 12, 2019, https://www.pewinternet.org/fact-sheet/mobile/

3   Aaron Smith, "Americans' Experiences with Data Security," Pew Research Center, January 26, 2017, https://www.pewinternet.org/2017/01/26/1-americans-experiences-with-data-security/

CHAPTER 1

1   See, e.g., Yuji Develle, "The Forgotten Origin of Passwords," *Medium*, September 1, 2016, https://medium.com/wonk-bridge/the-forgotten-origin-of-passwords-37c64427cd01; Gordon C. Aymar, *The Art of Portrait Painting* (Philadelphia: Chilton Book Company, 1967)

2   See, e.g., Elizabeth Loftus, *Eyewitness Testimony: With a New Preface* (Cambridge, MA: Harvard University Press, 1996)

3   Stephen Mayhew, "History of Biometrics," BiometricUpdate, https://www.biometricupdate.com/201802/history-of-biometrics-2 (accessed April 17, 2020)

4   Martijn van Mensvoort, "The History of Fingerprinting & the Study of Dermatoglyphics," Hand Research, http://fingerprints.handresearch.com/dermatoglyphics/fingerprints-history.htm (accessed April 17, 2020)

5   Van Mensvoort, "The History of Fingerprinting & the Study of Dermatoglyphics"

6   "The Bertillon System," New York Division of Criminal Justice Services, https://www.criminaljustice.ny.gov/ojis/history/bert_sys.htm (accessed April 17, 2020)

7   Matthew Wills, "Fingerprints and Crime," *Daily JSTOR*, June 8, 2018, https://daily.jstor.org/fingerprints-and-crime/

8   Mayhew, "History of Biometrics"

9   United States Federal Bureau of Investigation, "The Identification Division of the Federal Bureau of Investigation: A Brief Outline of the History, the Services, and the Operating Techniques of the World's Greatest Repository of Fingerprints," Federal Bureau of Investigation, US Department of Justice, 1977, https://catalog.hathitrust.org/Record/003787151

10   Arthur Conan Doyle, "A Study in Scarlet," Project Gutenberg, July 12, 2008, https://www.gutenberg.org/files/244/244-h/244-h.htm

11   Todd C. Pataky, Tingting Mu, Kerstin Bosch, Dieter Rosenbaum, and John Y. Goulermas, "Gait Recognition: Highly Unique Dynamic Plantar Pressure Patterns Among 104 Individuals," *Journal of the Royal Society Interface*, September 7, 2011, https://royalsocietypublishing.org/doi/full/10.1098/rsif.2011.0430?sid=a8ee14c0-0d7a-470f-b95b-1a9046e4d2ae

12   Bruce Schneier, "Gait Analysis from Satellite," Schneier on Security, September 9, 2008, https://www.schneier.com/blog/archives/2008/09/gait_analysis_f.html

13   Brad Kelechava, "Your Walk Shows Who You Are," ANSI, May 20, 2018, https://blog.ansi.org/2018/05/gait-analysis-walk-biometric-identification/

14   Alex Perala, "Researchers Say Gait Recognition Systems Offers 99.3% Accuracy," Find Biometrics, May 28, 2018, https://findbiometrics.com/researchers-gait-recognition-system-505286/

15   Dake Kang, "Chinese 'Gait Recognition' Tech IDs People by How They Walk," *AP News*, November 6, 2018, https://apnews.com/bf75dd1c26c947b7826d270a16e2658a

16   Yumiko Otsuka, "Face Recognition in Infants: A Review of Behavioral and Near-Spectroscopic Studies," *Japanese Psychological Research* 56, no. 1 (2014), 76–90, https://onlinelibrary.wiley.com/doi/pdf/10.1111/jpr.12024

17   Jennifer Tucker, "How Facial Recognition Technology Came to Be," *Boston Globe*, November 23, 2014, https://www.bostonglobe.com/ideas/2014/11/23/facial-recognition-technology-goes-way-back/CkWaxzozvFcveQ7kvdLHGI/story.html

18   Tucker, "How Facial Recognition Technology Came to Be"

19   Michael Ballantyne, Robert S. Boyer, and Larry Hines, "Woody Bledsoe: His Life and Legacy," *AI Magazine* 17, no. 1 (1996), 7, https://www.aaai.org/ojs/index.php/aimagazine/article/view/1207

20   Mayhew, "History of Biometrics"

21   Neil Lydick, "A Brief Overview of Facial Recognition," Electrical Engineering and Computer Science (EECS) Department, University of Michigan, http://www.eecs.umich.edu/courses/eecs487/w07/sa/pdf/nlydick-facial-recognition.pdf (accessed April 17, 2020)

22   Jesse Davis West, "19 Expert Quotes About the Future of Face Recognition," FaceFirst, April 16, 2018, https://www.facefirst.com/blog/quotes-about-the-future-of-face-recognition/

23   Martin Zizi, "The Flaws and Dangers of Facial Recognition," *Security Today*, March 1, 2019, https://securitytoday.com/articles/2019/03/01/the-flaws-and-dangers-of-facial-recognition.aspx

24   See, e.g., Steve Lohr, "Facial Recognition Is Accurate, If You're a White Guy," *New York Times*, February 9, 2018, https://www.nytimes.com/2018/02/09/technology/facial-recognition-race-artificial-intelligence.html

25   "Privacy Is Under Threat from the Facial Recognition Revolution," *Financial Times*, October 3, 2017, https://www.ft.com/content/4707f246-a760-11e7-93c5-648314d2c72c. See also Rodrigo

Salvaterra, "Demand Curves and the Dangers of AI: How AI Could Reveal Customers' Willingness to Pay Using Surveillance Cameras in Retail Stores," *Towards Data Science,* March 29, 2019, https://towardsdatascience.com/demand-curves-and-the-dangers-of-ai-65233bd77da8

26 See, e.g., the Illinois Biometric Information Protection Act and *Patel v. Facebook* case, both discussed in a later chapter of this book. See also, Tanya Harrington, "2020's Biggest Beauty Trend: Facial Un-Recognition," *Medium*, January 25, 2020, https://medium.com/@tanyaharrington96/2020s-biggest-beauty-trend-facial-un-recognition-1e696e32e241; Maya Lothian-McQueen, "These Activists Use Makeup to Defy Mass Surveillance," *i-D*, January 29, 2020, https://i-d.vice.com/en_uk/article/jge5jg/dazzle-club-surveillance-activists-makeup-marches-london-interview; and Aaron Holmes, "These Clothes Use Outlandish Designs to Trick Facial Recognition Software into Thinking You're Not Human," *Business Insider*, January 17, 2020, https://www.businessinsider.com/clothes-accessories-that-outsmart-facial-recognition-tech-2019-10

27 Katherine Schwab, "You're Already Being Watched by Facial Recognition Tech. This Map Shows Where," *Fast Company,* July 23, 2019, https://www.fastcompany.com/90379969/youre-already-being-watched-by-facial-recognition-tech-this-map-shows-where

28 See, e.g., Kathleen Foody, "Facebook Agrees to $550M BIPA Settlement," *Chicago Daily Law Bulletin*, February 10, 2020, https://www.chicagolawbulletin.com/facebook-agrees-to-$550m-bipa-settlement-20200210

29 Janine Jackson, "'Face Surveillance Is a Uniquely Dangerous Technology,'" FAIR, February 5, 2019, https://fair.org/home/face-surveillance-is-a-uniquely-dangerous-technology/

30 Naomi Kresge, Ilya Khrennikov, and David Ramli, "Period-Tracking Apps Are Monetizing Women's Extremely Personal Data," *Bloomberg Businessweek*, January 24, 2019, https://www.bloomberg.com/news/articles/2019-01-24/how-period-tracking-apps-are-monetizing-women-s-extremely-personal-data

31 Apps like 7Cups (https://www.7cups.com/) connect users to trained listeners, while platforms like Wysa (https://www.wysa.io/) connect users with a combination of therapists and AI chatbots.

32 "Human Genome Project Timeline of Events," National Human Genome Research Institute, https://www.genome.gov/human-genome-project/Timeline-of-Events (accessed April 17, 2020)

33 "Human Genome Project FAQ," National Human Genome Research Institute, https://www.genome.gov/human-genome-project/Completion-FAQ (accessed April 17, 2020)

34 See, e.g., "Celera: A Unique Approach to Genome Sequencing," Biocomputing Department, UC Berkeley, https://www.ocf.berkeley.edu/~edy/genome/celera.html (accessed April 17, 2020)

35 Thomas Fuller and Christine Hauser, "Search for 'Golden State Killer' Leads to Arrest of Ex-Cop," *New York Times*, April 25, 2018, https://www.nytimes.com/2018/04/25/us/golden-state-killer-serial.html

36 Wikipedia contributors, "Golden State Killer," Wikipedia, https://en.wikipedia.org/wiki/Golden_State_Killer (accessed April 17, 2020)

37 Heather Murphy and Tim Arango, "Joseph DeAngelo Pleads Guilty in Golden State Killer Cases," *New York Times*, June 29, 2020, https://www.nytimes.com/2020/06/29/us/golden-state-killer-joseph-deangelo.html

38 Nico Neumann, Catherine E. Tucker, and Timothy Whitfield, "How Effective Is Third-Party Consumer Profiling and Audience Delivery? Evidence from Field Studies," Forthcoming in *Marketing Science: Frontiers,* July 16, 2018, https://papers.ssrn.com/sol3/papers .cfm?abstract_id=3203131

39 Virginia Heffernan, "Just Google It: A Short History of a Newfound Verb," *Wired,* November 15, 2017, https://www.wired.com/story/just-google-it-a-short-history-of-a-newfound-verb/

40 Henry Ajder, Giorgio Patrini, Francesco Cavalli, and Laurence Cullen, "The State of Deepfakes: Landscape, Threats, and Impact," Deeptrace, September 2019, https://storage.googleapis.com /deeptrace-public/Deeptrace-the-State-of-Deepfakes-2019.pdf

41 Robert Chesney and Danielle Citron, "Deep Fakes: A Looming Challenge for Privacy, Democracy, and National Security," *University of California Law Review* 1753 (2019), July 21, 2018, https://papers.ssrn.com/sol3/papers.cfm?abstract_id=3213954##

42 Ajder et al., "The State of Deepfakes"

43 See Ajder et al., "The State of Deepfakes"; Oscar Schwartz, "You Thought Fake News Was Bad?", *Guardian,* November 12, 2018, https://www.theguardian.com/technology/2018/nov/12 /deep-fakes-fake-news-truth; and Lisa Vaas, "Deepfakes Have Doubled, Overwhelmingly Targeting Women," *Naked Security,* October 9, 2019, https://nakedsecurity.sophos.com/2019/10/09 /deepfakes-have-doubled-overwhelmingly-targeting-women/

44 Craig Silverman, "How to Spot a Deepfake Like the Barack Obama–Jordan Peele Video," *Buzz-Feed,* April 17, 2018, https://www.buzzfeed.com/craigsilverman/obama-jordan-peele-deepfake -video-debunk-buzzfeed

45 Bloomberg, "How Faking Videos Became So Easy – and Why That's So Scary," *Fortune,* September 11, 2018, https://fortune.com/2018/09/11/deep-fakes-obama-video/

46 Ajder et al., "The State of Deepfakes"

47 Samantha Cole, "We Are Truly Fucked: Everyone Is Making AI-Generated Fake Porn Now," *Vice,* January 24, 2018, https://www.vice.com/en_us/article/bjye8a/reddit-fake-porn-app-daisy -ridley

48 Chesney and Citron, "Deep Fakes"

49 Chesney and Citron, "Deep Fakes"

50 Ajder et al., "The State of Deepfakes"

51 See, e.g., Silverman, "Hot to Spot a Deepfake" (mouths); and Siwei Lyu, "Detecting 'Deepfake' Videos in the Blink of an Eye," PRI, September 5, 2018, https://www.pri.org/stories/2018-09-05 /detecting-deepfake-videos-blink-eye (eyes)

52 Kevin Stankiewicz, "'Perfectly Real' Deepfakes Will Arrive in 6 Months to a Year, Technology Pioneer Hao Li Says," CNBC, September 20, 2019, https://www.cnbc.com/2019/09/20/hao -li-perfectly-real-deepfakes-will-arrive-in-6-months-to-a-year.html

CHAPTER 2

1 Darren Orf, "Moore's Law Turns 50, and Everything Else You Missed This Weekend," *Gizmodo,* April 20, 2015, https://gizmodo.com/moores-law-turns-50-and-everything-else-you-missed -thi-1698888006

2   Gordon E. Moore, "Cramming More Components onto Integrated Circuits," *Electronics*, April 1965, http://www.cs.utexas.edu/~fussell/courses/cs352h/papers/moore.pdf (accessed April 17, 2020)

3   Moore, "Cramming More Components onto Integrated Circuits"

4   Moore, "Cramming More Components onto Integrated Circuits"

5   Maddie Stone, "The Trillion Fold Increase in Computing Power, Visualized," *Gizmodo*, May 24, 2015, https://gizmodo.com/the-trillion-fold-increase-in-computing-power-visualiz-1706676799

6   Bernard Marr, "How Much Data Do We Create Every Day? The Mind-Blowing Stats Everyone Should Read," *Forbes*, May 21, 2018, https://www.forbes.com/sites/bernardmarr/2018/05/21/how-much-data-do-we-create-every-day-the-mind-blowing-stats-everyone-should-read/#70694df660ba

7   Marr, "How Much Data Do We Create Every Day?"

8   Marr, "How Much Data Do We Create Every Day?"; Jeff Desjardins, "How Much Data Is Generated Each Day?" *Visual Capitalist*, April 15, 2019, https://www.visualcapitalist.com/how-much-data-is-generated-each-day/

9   Bernard Marr, "Big Data: Twenty Mind-Boggling Facts Everyone Must Read," *Forbes*, September, 30, 2015, https://www.forbes.com/sites/bernardmarr/2015/09/30/big-data-20-mind-boggling-facts-everyone-must-read/#1ae6efba17b1

10 Marr, "How Much Data Do We Create Every Day?"

11 Desjardins, "How Much Data Is Generated Each Day?"

12 Desjardins, "How Much Data Is Generated Each Day?"

13 Todd Hoff, "How Big Is a Petabyte, Exabyte, Zettabyte, or a Yottabyte?" *High Scalability*, September 11, 2012, http://highscalability.com/blog/2012/9/11/how-big-is-a-petabyte-exabyte-zettabyte-or-a-yottabyte.html

14 Henry Winchester, "A Brief History of Wearable Tech," *Wareable*, May 6, 2015, https://www.wareable.com/wearable-tech/a-brief-history-of-wearables

15 Sophie Charara, "245 Million Wearable Devices Will Be Sold in 2019," *Wareable*, September 1, 2015, https://www.wareable.com/wearable-tech/245-million-wearable-devices-sold-2019-1606

CHAPTER 3

1   The absence of an explicitly articulated right to privacy has been a longstanding complaint of abortion opponents hoping to see the Supreme Court overrule the precedent that was established in *Roe v. Wade*. *Roe* rested on a Constitutional right to privacy that was articulated by the Supreme Court in the landmark case *Griswold v. Connecticut*, finding that individuals had a right to privacy in their personal lives that extended to a right to access birth control. This was a groundbreaking case at the time, both because it overturned decades of laws restricting access to contraception, and because of the Constitutional right to privacy that the Court's opinion described.

2   E. L. Godkin, "Libel and Its Legal Remedy," *Journal of Social Science: Containing the Transactions of the American Association*, 12 (1880), 69

3   John H. De May and Alfred B. Scattergood v. Alvira Roberts, 46 Mich. 160; 9 N. W. 146 (Mich. 1881), https://faculty.uml.edu/sgallagher/DeMay.htm (accessed April 17, 2020)

4 Samuel Warren and Louis Brandeis, "The Right to Privacy," *Harvard Law Review* 4, no. 5 (December 15, 1890), 193–220, https://www.jstor.org/stable/1321160

5 Warren and Brandeis, "The Right to Privacy"

6 Warren and Brandeis, "The Right to Privacy"

7 "Restatement of the Law, Second, Torts, § 652," The American Law Institute, 1977, https://cyber .harvard.edu/privacy/Privacy_R2d_Torts_Sections.htm (accessed April 17, 2020)

8 "Restatement of the Law, Second, Torts, § 652"

9 "False Light," Digital Media Law Project, http://www.dmlp.org/legal-guide/false-light (accessed April 17, 2020)

10 "Restatement of the Law, Second, Torts, § 652"

11 "Restatement of the Law, Second, Torts, § 652"

12 Olmstead v. US, 277 US 438, 464–465 (1928), https://supreme.justia.com/cases/federal/us /277/438/#tab-opinion-1932307 (accessed April 17, 2020)

13 Warren and Brandeis, "The Right to Privacy"

14 De May and Scattergood v. Roberts, 46 Mich. 160; 9 N.W. 146 (Mich. 1881)

CHAPTER 4

1 Alessandro Acquisiti, Leslie John, and George Lowenstein, "What Is Privacy Worth?" Heinz College, Carnegie Mellon University, https://www.heinz.cmu.edu/~acquisti/papers/acquisti-ISR -worth.pdf (accessed April 17, 2020)

2 See also, Alessandro Acquisiti and Jenn Grossklags, "What Can Behavioral Economics Teach Us About Privacy," December 2007, https://www.researchgate.net/publication/228628228_What_ Can_Behavioral_Economics_Teach_Us_about_Privacy; Alessandro Acquisiti, Leslie K. John, and George Lowenstein, "The Impact of Relative Standards on the Propensity to Disclose," *Journal of Marketing Research* 49, no. 2 (April 1, 2012), https://journals.sagepub.com/doi/10.1509 /jmr.09.0215; and Alessandro Acquisiti, Leslie K. John, and George Loewenstein, "Strangers on a Plane: Context-Dependent Willingness to Disclose Sensitive Information," https://www.cmu .edu/dietrich/sds/docs/loewenstein/StrangersPlane.pdf

3 Acquisiti and Grossklags, "What Can Behavioral Economics Teach Us About Privacy"

4 Acquisiti, John, and Lowenstein, "What Is Privacy Worth?"

5 Acquisiti, John, and Lowenstein, "What Is Privacy Worth?"

6 Acquisiti, John, and Lowenstein, "What Is Privacy Worth?"

7 Ylan Mui, "Would You Give Up Google for $17,000 a Year? The Fed Wants to Know," CNBC, October 11, 2019, https://www.cnbc.com/2019/10/11/fed-tries-to-figure-out-value-of-free-internet -services-to-americans.html

8 See, e.g., David Byrne and Carol Corrado, "Accounting for Innovations in Consumer Digital Services: IT Still Matters," Finance and Economics Discussion Series, Divisions of Research & Statistics and Monetary Affairs, Federal Reserve Board, Washington, DC, 2019-049, https:// www.federalreserve.gov/econres/feds/files/2019049pap.pdf; and Erik Brynjolfsson, Avinash

Collis, and Felix Eggers, "Using Massive Online Choice Experiments to Measure Changes in Well-Being," *PNAS*, April 9, 2019, https://www.pnas.org/content/116/15/7250

9 Brynjolfsson, Collis, and Eggers, "Using Massive Online Choice Experiments to Measure Changes in Well-Being"

10 Brynjolfsson, Collis, and Eggers, "Using Massive Online Choice Experiments to Measure Changes in Well-Being"

11 Brynjolfsson, Collis, and Eggers, "Using Massive Online Choice Experiments to Measure Changes in Well-Being"

12 Brynjolfsson, Collis, and Eggers, "Using Massive Online Choice Experiments to Measure Changes in Well-Being"

13 "Digital Platforms Inquiry, Preliminary Report," Australian Competition & Consumer Commission, December 2018, 164, https://www.accc.gov.au/focus-areas/inquiries/digital-platforms-inquiry (accessed April 17, 2020)

14 "Digital Platforms Inquiry, Preliminary Report," 174

15 "Guide Concerning the Use of the Word 'Free' and Similar Representations," Federal Trade Commission, 16 CFR Part 251, https://www.ftc.gov/enforcement/rules/rulemaking-regulatory-reform-proceedings/guide-concerning-use-word-free-similar (accessed April 17, 2020)

SECTION II

1 "The Top 20 Valuable Facebook Statistics – Updated April 2020," Zephoria, https://zephoria.com/top-15-valuable-facebook-statistics/ (accessed April 17, 2020)

2 Elena Holodny, "Computers Using Facebook 'Likes' May Be Assessing You More Accurately Than Your Friends – and Researchers Warned This Could Be Misused," *Business Insider*, September 29, 2017, https://www.businessinsider.com/what-facebook-likes-say-about-you-2017-9

3 Wu Youyou, Michal Kosinski, and David Stillwell, "Computer-Based Personality Judgments Are More Accurate Than Those Made by Humans," *PNAS* 112, no. 4 (January 27, 2015), 1036–1040, https://www.pnas.org/content/pnas/112/4/1036.full.pdf

4 "The requirement for announcing the intention to create or enlarge a system stems from our conviction that public involvement is essential for fully effective consideration of the pros and cons of establishing a personal data system. Opportunity for public involvement must not be limited to actual or potential data subjects; it should extend to all individuals and interests that may have views on the desirability of a system." "Records, Computers, and the Rights of Citizens," HEW Report, Sec. IV, "Recommended Safeguards for Administrative Personal Data Systems," subsection II, "Public Notice Requirement," https://aspe.hhs.gov/report/records-computers-and-rights-citizens (accessed April 17, 2020)

5 See, e.g., "The Openness Principle may be viewed as a prerequisite for the Individual Participation Principles; for the latter principle to be effective, it must be possible in practice to acquire information about the collection, storage, or use of personal data. Regular information from data controllers on a voluntary basis, publication in official registers of descriptions of activities concerned with the processing of personal data, and registration

with public bodies are some, though not all, of the ways by which this may be brought about." Detailed comments to paragraph 12, "OECD Guidelines on the Protection of Privacy and Transborder Flows of Personal Data," 1980, http://www.oecd.org/sti/ieconomy/oecdguidelinesontheprotectionofprivacyandtransborderflowsofpersonaldata.htm (accessed April 17, 2020)

6 "Specifically, the notice-and-choice model, as implemented, has led to long, incomprehensible privacy policies that consumers typically do not read, let alone understand. Likewise, the harm-based model has been criticized for failing to recognize a wider range of privacy-related concerns, including reputational harm or the fear of being monitored." "Protecting Consumer Privacy in an Era of Rapid Change: A Proposed Framework for Businesses and Policymakers," Preliminary FTC Staff Report, Federal Trade Commission, December 2010, iii, https://www.ftc.gov/sites/default/files/documents/reports/federal-trade-commission-bureau-consumer-protection-preliminary-ftc-staff-report-protecting-consumer/101201privacyreport.pdf

7 "Biometric Information Privacy Act," Illinois General Assembly, P.A. 95-994, eff. 10-3-08, http://www.ilga.gov/legislation/ilcs/ilcs3.asp?ActID=3004 (accessed April 17, 2020)

8 Rosenbach v. Six Flags Entertainment Corp., 2019 IL 123186, https://courts.illinois.gov/Opinions/SupremeCourt/2019/123186.pdf (accessed April 17, 2020) because the user posted it publicly

9 Seyfarth Shaw, "Copy-Cat Class Actions Meet Copy-Cat Legislation: Illinois' BIPA Spurs New Biometric Privacy Legislation Across the Nation," *Workplace Class Action Blog*, July 11, 2019, https://www.workplaceclassaction.com/2019/07/copy-cat-class-actions-meet-copy-cat-legislation-illinois-bipa-spurs-new-biometric-privacy-legislation-across-the-nation/

10 "California Consumer Privacy Act," California Legislative Information, June 29, 2018, https://leginfo.legislature.ca.gov/faces/billTextClient.xhtml?bill_id=201720180AB375

11 California SOS Press Office, "Secretary of State Alex Padilla Certifies Measures for the November 3, 2020 General Election Ballot," news release no. AP20:059, June 25, 2020, https://www.sos.ca.gov/administration/news-releases-and-advisories/2020-news-releases-and-advisories/ap20059-secretary-state-alex-padilla-certifies-measures-november-3-2020-general-election-ballot/

CHAPTER 5

1 Duncan McCann and Miranda Hall, "Blocking the Data Stalkers: Going Beyond GDPR to Tackle Power in the Data Economy," New Economics Foundation, December 20, 2018, 4, https://neweconomics.org/2018/12/blocking-the-data-stalkers

2 McCann and Hall, "Blocking the Data Stalkers," 7

3 "FTC Fact Sheet: Antitrust Laws: A Brief History," Federal Trade Commission, https://www.consumer.ftc.gov/sites/default/files/games/off-site/youarehere/pages/pdf/FTC-Competition_Antitrust-Laws.pdf (accessed April 17, 2020)

4 See, e.g., Elizabeth Warren, "Here's How We Can Break Up Big Tech," *Medium*, March 8, 2019, https://medium.com/@teamwarren/heres-how-we-can-break-up-big-tech-9ad9e0da324c

5 "Digital Platforms Inquiry, Preliminary Report," 166–172

6 "Digital Platforms Inquiry, Preliminary Report," 186

7 Alexis C. Madrigal, "Reading the Privacy Policies You Encounter in a Year Would Take 76 Work Days," *Atlantic*, March 1, 2012, https://www.theatlantic.com/technology/archive/2012/03 /reading-the-privacy-policies-you-encounter-in-a-year-would-take-76-work-days/253851/

8 In re: Facebook, Inc., Consumer Privacy User Profile Litigation, MDL No. 2843 Case No. 18-md-02843-VC, Pretrial Order No. 20, Granting in Part and Denying in Part Defendant's Motion to Dismiss First Amended Complaint, September 9, 2019, 21, https://cand.uscourts.gov /vc/fbmdl

9 In re: Facebook, Inc., Consumer Privacy User Profile Litigation, Pretrial Order No. 20, at p. 25–26

10 "Digital Platforms Inquiry, Preliminary Report," 186

11 Caroline Knorr, "The Sneaky Science Behind Your Child's Tech Obsession," *Washington Post*, November 9, 2018, https://www.washingtonpost.com/lifestyle/2018/11/09/sneaky-science -behind-your-childs-tech-obsession/?utm_term=.ba5d24e0d535

12 Knorr, "The Sneaky Science Behind Your Child's Tech Obsession"

13 "Digital Platforms Inquiry, Preliminary Report," 187

14 Sam Schechner and Mark Secada, "You Give Apps Sensitive Personal Information. Then They Tell Facebook," *Wall Street Journal*, February 22, 2019, https://www.wsj.com/articles/you-give -apps-sensitive-personal-information-then-they-tell-facebook-11550851636?mod=e2tw

15 Nick Statt, "App Makers Are Sharing Sensitive Personal Information with Facebook but Not Telling Users," *Verge,* February 22, 2019, https://www.theverge.com/2019/2/22/18236398 /facebook-mobile-apps-data-sharing-ads-health-fitness-privacy-violation

16 Wolfie Christl, "Corporate Surveillance in Everyday Life: How Companies Collect, Combine, Analyze, Trade, and Use Personal Data on Billions," *Cracked Labs*, June 2017, 4, https:// crackedlabs.org/en/corporate-surveillance

17 Geoffrey Fowler, "It's the Middle of the Night. Do You Know Who Your iPhone Is Talking To?" *Washington Post*, May 28, 2019, https://www.washingtonpost.com/technology/2019/05/28 /its-middle-night-do-you-know-who-your-iphone-is-talking/?utm_term=.6b32a8991a36

18 Fowler, "It's the Middle of the Night. Do You Know Who Your iPhone Is Talking To?"

19 See, e.g., Pierre de Poulpiquet, "What Is a Walled Garden of Data, and Why Is It the Strategy of Google, Facebook, and Amazon Ads Platforms?" *Medium*, November 3, 2017, https://medium .com/mediarithmics-what-is/what-is-a-walled-garden-and-why-it-is-the-strategy-of-google -facebook-and-amazon-ads-platform-296ddeb784b1

20 Christl, "Corporate Surveillance in Everyday Life," 40

21 "Data Brokers: A Call for Transparency and Accountability," Federal Trade Commission, May 2014, iv, https://www.ftc.gov/system/files/documents/reports/data-brokers-call-transparency -accountability-report-federal-trade-commission-may-2014/140527databrokerreport.pdf

22 Christl, "Corporate Surveillance in Everyday Life," Section 5.1; and "Data Brokers: A Call for Transparency and Accountability," iv–v

23 "Data Brokers: A Call for Transparency and Accountability," ii–iii

24 Christl, "Corporate Surveillance in Everyday Life," Section 5.1

25 "Data Brokers: A Call for Transparency and Accountability"

26 Devin Coldewey, "Vermont Passes First Law to Crack Down on Data Brokers," TechCrunch, May 27, 2018, https://techcrunch.com/2018/05/27/vermont-passes-first-first-law-to-crack-down-on-data-brokers/

27 "California Consumer Privacy Act"

28 "Digital Platforms Inquiry, Preliminary Report," 193

29 Cecilia Kang, "FTC Approves Facebook Fine of About $5 Billion," *New York Times,* July 12, 2019, https://www.nytimes.com/2019/07/12/technology/facebook-ftc-fine.html

30 Kang, "FTC Approves Facebook Fine of About $5 Billion"

31 Christl, "Corporate Surveillance in Everyday Life"

32 David Meyer, "The Privacy and Antitrust Worlds Are Starting to Cross Over," IAPP Privacy Advisor, April 23, 2019, https://iapp.org/news/a/the-privacy-and-antitrust-worlds-are-starting-to-cross-over/

33 "Consumer Protection in the States: A 50-State Evaluation of Unfair and Deceptive Practices Laws," National Consumer Law Center, March 2018, https://www.nclc.org/issues/how-well-do-states-protect-consumers.html

34 Letter from Attorneys General for California, Connecticut, District of Columbia, Illinois, Massachusetts, Minnesota, Mississippi, New York, Oregon, Pennsylvania, Rhode Island, and Washington, to the Federal Trade Commission, "Re: Competition and Consumer Protection in the 21st Century Hearing Project No. P181201 Antitrust/ Competition Issues," October 10, 2018, https://oag.ca.gov/system/files/attachments/press-docs/10.10.2018-multistate-ag-letter-ftc-re-hearings.pdf

35 State AG letter to FTC, 3

36 State AG letter to FTC, 3

37 State AG letter to FTC, 3

38 State AG letter to FTC, 4

39 State AG letter to FTC, 8

40 State AG letter to FTC, 4

41 "Attorney General James Gives Update on Facebook Antitrust Investigation," New York State Office of the Attorney General, October 22, 2019, https://ag.ny.gov/press-release/2019/attorney-general-james-gives-update-facebook-antitrust-investigation

CHAPTER 6

1 Benjamin Wittes and Jodie Liu, "The Privacy Paradox: The Privacy Benefits of Privacy Threats," Brookings Institution, May 2015, https://www.brookings.edu/wp-content/uploads/2016/06/Wittes-and-Liu_Privacy-paradox_v10.pdf

2 "Our Story," Interactive Advertising Bureau (IAB), https://www.iab.com/our-story/ (accessed April 17, 2020)

3   See, e.g., Maciej Zawadziński, "How Does Real-Time Bidding (RTB) Work?" ClearCode,
    https://clearcode.cc/blog/real-time-bidding/ (accessed April 17, 2020); and McCann and Hall,
    "Blocking the Data Stalkers," 9

4   Chris Ip, "An Early Test of the GDPR: Taking On the Data Brokers," *Engadget*, November 8,
    2018, https://www.engadget.com/2018/11/08/gdpr-data-brokers-complaints/; and Amit Katwala,
    "Forget Facebook, Mysterious Data Brokers Are Facing GDPR Trouble," *Wired*, November 8,
    2018, https://www.wired.co.uk/article/gdpr-acxiom-experian-privacy-international-data-brokers

5   "Why We've Filed Complaints Against Companies That Most People Have Never Heard Of –
    and What Needs to Happen Next," Privacy International, November 8, 2018, https://privacy
    international.org/advocacy/2434/why-weve-filed-complaints-against-companies-most-people
    -have-never-heard-and-what

6   Ryan Calo and Alex Rosenblatt, "The Taking Economy: Uber, Information, and Power," *Colum-
    bia Law Review* 117 (March 11, 2017), 1627–1628, https://papers.ssrn.com/sol3/papers
    .cfm?abstract_id=2929643

7   Calo and Rosenblatt, "The Taking Economy," 1628

8   Mike Isaac, "How Uber Deceives the Authorities Worldwide," *New York Times*, March 3, 2017,
    https://www.nytimes.com/2017/03/03/technology/uber-greyball-program-evade-authorities.html

9   The allegations of near-constant changes in contract terms, pricing and reimbursement, and
    variations in wait time calculations required to collect no-show fees, are more related to labor
    law and fair employment conditions than to data privacy concerns as such.

10 Calo and Rosenblatt, "The Taking Economy," 1629–1631

11 Calo and Rosenblatt, "The Taking Economy," 1652

12 Nicole Martin, "Uber Charges More If They Think You're Willing to Pay More," *Forbes*, March
    30, 2019, https://www.forbes.com/sites/nicolemartin1/2019/03/30/uber-charges-more-if-they
    -think-youre-willing-to-pay-more/

13 Calo and Rosenblatt, "The Taking Economy," 1658–1659

14 Geoffrey Fowler, "When Tax Prep Is Free, You May Be Paying with Your Privacy," *Washington
    Post,* March 7, 2019, https://www.washingtonpost.com/technology/2019/03/07/when-tax-prep-is
    -free-you-may-be-paying-with-your-privacy/?utm_term=.dfe428c19327

15 Rachel Kraus, "Finally, a Bipartisan Issue: Accepting Money from Lobbyists to Block Free
    Tax-Filing Software," *Mashable*, April 9, 2019, https://mashable.com/article/free-tax-filing
    -software/

16 Adam Walser, "Florida DMV Sells Your Personal Data to Private Companies, Marketing
    Firms," WPTV, July 11, 2019, https://www.wptv.com/news/state/florida-dmv-sells-your-personal
    -information-to-private-companies-marketing-firms

17 "Extensive Coverage. We activate data across an ecosystem of more than 575 partners, repre-
    senting the largest network of connections in the digital marketing space. We use 100% deter-
    ministic matching, resulting in the strongest combination of reach and accuracy. We offer
    multi-sourced insight into approximately 700 million consumers worldwide, and our data
    products contain over 5,000 data elements from hundreds of sources with permission rights . . .

Strong Client Relationships. We serve more than 2,500 clients directly . . ." Acxiom 2018 Annual Report, 12–13, http://www.annualreports.com/HostedData/AnnualReports/PDF/NASDAQ_ACXM_2018.pdf

18 https://www.acxiom.com/ (accessed April 17, 2020)

19 https://www.acxiom.com/

20 Acxiom 2018 Annual Report, 12

21 https://www.acxiom.com/

22 "Consumer Insights Packages," Acxiom, https://www.acxiom.com/what-we-do/data-packages/ (accessed April 17, 2020)

23 "Consumer Insights Packages"

24 "Consumer Insights Packages"

25 "2019 Data Directory," Oracle, http://www.oracle.com/us/solutions/cloud/data-directory -2810741.pdf (accessed April 17, 2020)

26 Natasha Lomas, "Covert Data-Scraping on Watch as EU DPA Lays Down 'Radical' GDPR Redline," TechCrunch, March 30, 2019, https://techcrunch.com/2019/03/30/covert-data-scraping -on-watch-as-eu-dpa-lays-down-radical-gdpr-red-line/; and Ewa Kurowska-Tober and Magdalena Koniarska, "Poland: Polish DPA Issues the First Fine for a Violation of the GDPR – and It's Harsh," DLA Piper, April 2, 2019, https://blogs.dlapiper.com/privacymatters/polish-dpa-issues -the-first-fine-for-a-violation-of-the-gdpr-and-its-harsh/

27 "Our Complaints Against Acxiom, Criteo, Equifax, Experian, Oracle, Quantcast, Tapad," Privacy International, November 8, 2018, https://privacyinternational.org/advocacy/2426/our -complaints-against-acxiom-criteo-equifax-experian-oracle-quantcast-tapad

CHAPTER 7

1 Wikipedia contributors, "The Minority Report," Wikipedia, https://en.wikipedia.org/wiki/The_ Minority_Report (accessed April 17, 2020)

2 "Computers Diagnose Parkinson's Disease with Behavioral Tracking," CleanroomConnect, https://cleanroomconnect.com/parkinsons-disease-computer-diagnosis/ (accessed April 17, 2020)

3 Jason Tashea, "Risk-Assessment Algorithms Challenged in Bail, Sentencing and Parole Decisions," ABA Journal, March 1, 2017, http://www.abajournal.com/magazine/article /algorithm_bail_sentencing_parole/

4 Joy Buolamwini, "Artificial Intelligence Has a Problem with Gender and Racial Bias. Here's How to Solve It," Time, February 7, 2019, https://time.com/5520558/artificial-intelligence-racial -gender-bias/

5 See, e.g., Bernard Marr, "What Is the Difference Between Artificial Intelligence and Machine Learning?" Forbes, December 6, 2016, https://www.forbes.com/sites/bernardmarr/2016/12/06 /what-is-the-difference-between-artificial-intelligence-and-machine-learning/#6d77702a2742

6 Marr, "What Is the Difference Between Artificial Intelligence and Machine Learning?"

7 Marr, "What Is the Difference Between Artificial Intelligence and Machine Learning?"

8 See, e.g., Charles McLellan, "Inside the Black Box: Understanding AI Decision-Making," ZDNet, December 1, 2016, https://www.zdnet.com/article/inside-the-black-box-understanding -ai-decision-making/

9 See, e.g., "AI Now Report 2018," AI Now Institute, New York University, December 2018, https://ainowinstitute.org/AI_Now_2018_Report.pdf

10 Tashea, "Risk-Assessment Algorithms Challenged in Bail, Sentencing and Parole Decisions"

11 Jeffrey Dastin, "Amazon Scraps Secret AI Recruiting Tool That Showed Bias Against Women," *Reuters*, October 9, 2018, https://www.reuters.com/article/us-amazon-com-jobs-automation -insight-idUSKCN1MK08G

12 "AI Now 2018 Report," 18

13 Cameron Langford, "Houston Schools Must Face Teacher Evaluation Lawsuit," *Courthouse News Service*, May 8, 2017, https://www.courthousenews.com/houston-schools-must-face-teacher -evaluation-lawsuit/

14 "AI Now 2018 Report"

15 Hope Reese, "Why Microsoft's 'Tay' AI Bot Went Wrong," *TechRepublic*, March 24, 2016, https://www.techrepublic.com/article/why-microsofts-tay-ai-bot-went-wrong/

16 Alice Gregory, "R U There?" *New Yorker,* February 2, 2015, https://www.newyorker.com /magazine/2015/02/09/r-u

17 https://www.crisistextline.org/

18 Gregory, "R U There?"

19 Gregory, "R U There?"

20 Kari Stephens, "How Can Mental Health Join in the Big Data movement?" *Psychiatry News*, May 1, 2017, https://psychnews.psychiatryonline.org/doi/full/10.1176/appi.pn.2017.5a15

21 Rebecca Ruiz, "AI Figured Out the Word People Text When Their Suicide Risk Is High," *Mashable*, March 9, 2017, https://mashable.com/2017/03/09/artificial-intelligence-suicide-risk/

22 Rebecca Ruiz, "The Crisis Text Line Analyzed 75 Million Texts to Pinpoint the Best Way to Ask If Someone's Suicidal," *Mashable*, January 30, 2019, https://mashable.com/article/how-to-ask -about-suicide-crisis-text-line/

23 James Vincent, "Facebook Is Using AI to Spot Users with Suicidal Thoughts and Send Them Help," *Verge*, November 28, 2017, https://www.theverge.com/2017/11/28/16709224/facebook -suicidal-thoughts-ai-help

24 "Board and Advisors," Crisis Text Line, https://www.crisistextline.org/board-advisors (accessed April 17, 2020)

25 Mike Isaac, "Facebook Offers Tools for Those Who Fear a Friend May Be Suicidal," *New York Times*, June 14, 2016, https://www.nytimes.com/2016/06/15/technology/facebook-offers-tools -for-those-who-fear-a-friend-may-be-suicidal.html?ref=technology&_r=2

26 Isaac, "Facebook Offers Tools for Those Who Fear a Friend May Be Suicidal"

27 Vindu Goel, "Facebook Promises Deeper Review of User Research, but Is Short on the Particulars," *New York Times*, October 2, 2014, https://www.nytimes.com/2014/10/03/technology /facebook-promises-a-deeper-review-of-its-user-research.html?module=inline

28 Goel, "Facebook Promises Deeper Review of User Research, but Is Short on the Particulars"

29 Vincent, "Facebook Is Using AI to Spot Users with Suicidal Thoughts and Send Them Help"

30 See., e.g., Hayley Tsukyama, "Facebook Is Using AI to Try to Prevent Suicide," *Washington Post*, November 27, 2017, https://www.washingtonpost.com/news/the-switch/wp/2017/11/27/facebook -is-using-ai-to-try-to-prevent-suicide/?utm_term=.d93c6e9b6d75; Isaac, "Facebook Offers Tools for Those Who Fear a Friend May Be Suicidal"; and Rebecca Ruiz, "Facebook's AI Suicide Prevention Tool Can Save Lives, but the Company Won't Say How It Works," *Mashable*, November 27, 2017, https://mashable.com/2017/11/28/facebook-ai-suicide-prevention-tools/

31 Ruiz, "Facebook's AI Suicide Prevention Tool Can Save Lives, but the Company Won't Say How It Works"

32 Sam T. Levin, "Facebook Told Advertisers It Can Identify Teens Feeling 'Insecure' and 'Worthless'," *Guardian,* May 1, 2017, https://www.theguardian.com/technology/2017/may/01 /facebook-advertising-data-insecure-teens

33 Lucy Tiven, "Facebook Announces Plan to Monitor Your Mental Health," *ATTN*, June 15, 2016, https://archive.attn.com/stories/9216/facebook-announces-plan-to-monitor-your-mental-health

CHAPTER 8

1   Craig Silverman, Jane Lytvynenko, and William Kung, "Disinformation for Hire: How a New Breed of PR Firms Is Selling Lies Online," *BuzzFeed*, January 6, 2020, https://www .buzzfeednews.com/article/craigsilverman/disinformation-for-hire-black-pr-firms

2   Douglas Rushkoff, "Why Mark Zuckerberg Thinks a President Elizabeth Warren Would 'Suck'," CNN, October 1, 2019, https://www.cnn.com/2019/10/01/opinions/mark -zuckerberg-facebook-elizabeth-warren-rushkoff/index.html

3   Special Counsel Robert S. Mueller, III, "Report on the Investigation into Russian Interference in the 2016 Presidential Election," US Department of Justice, volume I of II, March 2019, https://www.justice.gov/storage/report.pdf

4   Ben Collins and Joseph Fox, "Jenna Abrams, Russia's Clown Troll Princess, Duped the Mainstream Media and the World," *Daily Beast*, November 3, 2017, https://www.thedailybeast.com /jenna-abrams-russias-clown-troll-princess-duped-the-mainstream-media-and-the-world

5   Josephine Lukito and Chris Wells, "Most Major Outlets Have Used Russian Tweets as Sources for Partisan Opinion: Study," *Columbia Journalism Review*, March 8, 2018, https://www.cjr.org /analysis/tweets-russia-news.php

6   Mueller, "Report on the Investigation into Russian Interference in the 2016 Presidential Election," 15

7   Mueller, "Report on the Investigation into Russian Interference in the 2016 Presidential Election," 24–26

8   Adam Satariano, "Russia Sought to Use Social Media to Influence E.U. Vote, Report Finds," *New York Times*, June 14, 2019, https://www.nytimes.com/2019/06/14/business/eu-elections -russia-misinformation.html

9 Anna Kaplan, "Judge Tosses Russian Trolls' Free Speech Lawsuit Against Facebook," *Daily Beast*, July 20, 2019, https://www.thedailybeast.com/judge-tosses-russian-trolls-free-speech -lawsuit-against-facebook?ref=scroll

10 In re: Facebook, Inc., Consumer Privacy User Profile Litigation

11 In re: Facebook, Inc., Consumer Privacy User Profile Litigation, Pretrial Order No. 20, 9–10

12 Salvador Rodriguez, "Mark Zuckerberg: I Thought About Banning Political Ads from Facebook, but Decided Not To," CNBC, October 17, 2019, https://www.cnbc.com/2019/10/17 /mark-zuckerberg-says-he-wont-ban-political-ads-on-facebook.html

13 Issie Lapowsky, "The Top Political Advertiser on Facebook Is . . . Facebook," *Wired*, Octber 23, 2018, https://www.wired.com/story/top-political-advertiser-on-facebook-is-facebook/

14 "About Us – Mission Statement," Alliance for Securing Democracy, https://securingdemocracy .gmfus.org/about-us/ (accessed April 17, 2020)

15 Kathleen Hall Jamieson, *Cyberwar: How Russian Trolls Helped Elect a President: What We Don't, Can't, and Do Know* (New York: Oxford University Press, 2017)

16 Briony Swire-Thompson, Ullrich K. H. Ecker, Stephan Lewandowsky, and Adam J. Berinsky, "They Might Be a Liar but They're My Liar: Source Evaluation and the Prevalence of Misinformation," *Political Psychology* 41, no. 1 (2020), https://doi.org/10.1111/pops.12586

17 See, e.g., "How Attempts to Moderate Online Content Have Fallen Short," Annenberg Public Policy Center, University of Pennsylvania, June 17, 2019, https://www.annenbergpublicpolicy center.org/governments-platforms-fallen-short-trying-to-moderate-content-online/

18 Casey Newton, "People Older Than 65 Share the Most Fake News, New Study Finds," *Verge*, January 9, 2019, https://www.theverge.com/2019/1/9/18174631/old-people-fake-news-facebook -share-nyu-princeton

19 Eliza Mackintosh, "Finland Is Winning the War on Fake News. What It's Learned May Be Crucial to Western Democracy," CNN, August 2, 2019, https://www.cnn.com/interactive/2019/05 /europe/finland-fake-news-intl/

CHAPTER 9

1 The Complaint, *Arias v. Intermex,* was filed on May 5, 2015 and is available online at http://cdn .arstechnica.net/wp-content/uploads/2015/05/Intermexcomplaint.pdf

2 Complaint at p. 3, paragraph 7

3 Complaint, p. 3, paragraph 7

4 Complaint, p. 4, paragraphs 13–14

5 Elizabeth Austermuehle, "Monitoring Your Employees Through GPS: What Is Legal, and What Are Best Practices?" Greensfelder, February 18, 2016, http://www.greensfelder.com/business -risk-management-blog/monitoring-your-employees-through-gps-what-is-legal-and-what -are-best-practices

6 Lucy Clarke-Billings, "Psychologists Warn Constant Email Notifications Are 'Toxic Source of Stress'," *Telegraph*, January 2, 2016, https://www.telegraph.co.uk/news/2016/03/22/psychologists-warn-constant-email-notifications-are-toxic-source/

7 Chris Matyszczyk, "French Workers Win Right to Ignore Work Emails After Hours," *CNET*, January 1, 2017, https://www.cnet.com/news/french-employees-law-ignore-work-emails-after-hours/

8 Jim Edwards, "Brutal Conditions in Amazon's Warehouses Threaten to Ruin the Company's Image," *Business Insider*, August 5, 2013, https://www.businessinsider.com/brutal-conditions-in-amazons-warehouses-2013-8

9 Shannon Liao, "Amazon Warehouse Workers Skip Bathroom Breaks to Keep Their Jobs, Says Report," *Verge*, April 16, 2018, https://www.theverge.com/2018/4/16/17243026/amazon-warehouse-jobs-worker-conditions-bathroom-breaks

10 Daphne Howland, "Why Amazon's 'Big Brother' Warehouse Theft Surveillance Is a Big Mistake," Retail Dive, March 23, 2016, https://www.retaildive.com/news/why-amazons-big-brother-warehouse-theft-surveillance-is-a-big-mistake/415764/

11 Howland, "Why Amazon's 'Big Brother' Warehouse Theft Surveillance Is a Big Mistake"

12 Michael Sainato, "'Go Back to Work': Outcry over Deaths on Amazon's Warehouse Floor," *Guardian*, October 18, 2019, https://www.theguardian.com/technology/2019/oct/17/amazon-warehouse-worker-deaths

13 Heather Kelly, "Amazon's Idea for Employee-Tracking Wearables Raises Concerns," CNN Business, February 2, 2018, https://money.cnn.com/2018/02/02/technology/amazon-employee-tracker/index.html; and Edwards, "Brutal Conditions in Amazon's Warehouses Threaten to Ruin the Company's Image"

14 Edwards, "Brutal Conditions in Amazon's Warehouses Threaten to Ruin the Company's Image"

15 Jena McGregor, "Some Swedish Workers Are Getting Microchips Implanted in Their Hands," *Washington Post*, April 4, 2017, https://www.washingtonpost.com/news/on-leadership/wp/2017/04/04/some-swedish-workers-are-getting-microchips-implanted-in-their-hands/

16 "Would You Get 'Microchipped' at Work?" YouTube video, 2:36, posted by CNN Business, July 25, 2017, https://www.youtube.com/watch?v=M-gE8FbqHUU

17 See, e.g., Thomas Heath, "This Employee ID Badge Monitors and Listens to You at Work – Except in the Bathroom," *Washington Post*, September 7, 2016, https://www.washingtonpost.com/news/business/wp/2016/09/07/this-employee-badge-knows-not-only-where-you-are-but-whether-you-are-talking-to-your-co-workers/; and McGregor, "Some Swedish Workers Are Getting Microchips Implanted in Their Hands"

18 "Employer Health Benefits, 2019 Annual Survey," Kaiser Family Foundation, 198, http://files.kff.org/attachment/Report-Employer-Health-Benefits-Annual-Survey-2019

19 "Employer Health Benefits, 2019 Annual Survey," 195

20 Sally Wadyka, "Are Workplace Wellness Programs a Privacy Problem?" *Consumer Reports*, January 16, 2020, https://www.consumerreports.org/health-privacy/are-workplace-wellness-programs-a-privacy-problem/

21 "Guidelines 3/2019 on Processing of Personal Data Through Video Devices," European Data Protection Board, July 10, 2019, https://edpb.europa.eu/sites/edpb/files/consultation/edpb_guidelines_201903_videosurveillance.pdf

22 Leonid Bershidsky, "Why the Modern Workplace Needs Punch Clocks," *Bloomberg*, May 15, 2019, https://www.bloomberg.com/opinion/articles/2019-05-16/eu-time-tracking-ruling-could-generate-useful-productivity-data; and In Case C-55/18, Judgment of the Court, Federación de Servicios de Comisiones Obreras (CCOO), May 14, 2019, http://curia.europa.eu/juris/document/document.jsf;jsessionid=8391DB149E1E5819119CBB5E0976C433?text=&docid=214043&pageIndex=0&doclang=EN&mode=req&dir=&occ=first&part=1&cid=3604643

23 Bershidsky, "Why the Modern Workplace Needs Punch Clocks"

24 In Case C-55/18, Judgment of the Court

25 David Frank, "AARP Wins Victory for Workers' Civil Rights," *AARP*, December 22, 2017, https://www.aarp.org/politics-society/advocacy/info-2017/eeoc-workers-rights-fd.html

26 Julie Appleby, "Workplace Wellness Programs Offer Big Incentives, But May Cost Your Privacy," NPR, September 22, 2018, https://www.npr.org/sections/health-shots/2018/09/22/649664555/workplace-wellness-plans-offer-big-incentives-but-may-cost-your-privacy

27 Ifeoma Ajunwa, "Workplace Wellness Programs Could Be Putting Your Health Data at Risk," *Harvard Business Review*, January 19, 2017, https://hbr.org/2017/01/workplace-wellness-programs-could-be-putting-your-health-data-at-risk

28 "About the Coalition for Genetic Fairness," Coalition for Genetic Fairness,http://www.geneticfairness.org/about.html (accessed April 17, 2020)

29 "Genetic Information Discrimination," US Equal Employment Opportunity Commission, https://www.eeoc.gov/laws/types/genetic.cfm (accessed April 17, 2020)

30 The bill, titled "Preserving Employee Wellness Programs Act (PEWPA), was introduced as H.R. 1313, available at https://www.congress.gov/bill/115th-congress/house-bill/1313/text

CHAPTER 10

1 Benjamin Herold, "Google Experimenting with New Cloud Storage, Artificial Intelligence Initiative for K-12," *Education Week*, April 4, 2019, https://blogs.edweek.org/edweek/DigitalEducation/2019/04/google_cloud_storage_artificial_intelligence_K-12.html

2 Herold, "Google Experimenting with New Cloud Storage, Artificial Intelligence Initiative for K-12"

3 Matt Day, "Amazon in the Classroom? Amazon's Voice Assistant Leads Story Time," *Los Angeles Times*, June 14, 2019, https://www.latimes.com/business/la-fi-amazon-alexa-classroom-20190614-story.html

4 See Children's Online Privacy Protection Rule ("COPPA"), available at https://www.ftc.gov/enforcement/rules/rulemaking-regulatory-reform-proceedings/childrens-online-privacy-protection-rule, and https://www.echokidsprivacy.com/

5 https://www.echokidsprivacy.com/

6 Priya C. Kumar, Marshini Chetty, Tamara L. Clegg, and Jessica Vitak, "Privacy and Security Considerations for Digital Technology Use in Elementary Schools," CHI 2019 Paper, May 4–9, 2019, https://hci.princeton.edu/wp-content/uploads/sites/459/2019/01/paper307.pdf

7 Kumar et al., "Privacy and Security Considerations for Digital Technology Use in Elementary Schools"

8 Camila Domonske, "SC Mom Says Baby Monitor Was Hacked; Experts Say Many Devices Are Vulnerable," NPR, June 5, 2018, https://www.npr.org/sections/thetwoway/2018/06/05/617196788 /s-c-mom-says-baby-monitor-was-hacked-experts-say-many-devices-are-vulnerable

9 Domonske, "SC Mom Says Baby Monitor Was Hacked; Experts Say Many Devices Are Vulnerable"

10 Domonske, "SC Mom Says Baby Monitor Was Hacked; Experts Say Many Devices Are Vulnerable"

11 Philip Olterman, "German Parents Told to Destroy Doll That Can Spy on Children," *Guardian*, February 17, 2017, https://www.theguardian.com/world/2017/feb/17/german-parents-told-to -destroy-my-friend-cayla-doll-spy-on-children

12 Olterman, "German Parents Told to Destroy Doll That Can Spy on Children"

13 Olterman, "German Parents Told to Destroy Doll That Can Spy on Children"

14 Diane Shipley, "Reminder: Smart Toys Are Cute, Cuddly, and Full of Security Risks," *Mashable*, January 26, 2019, https://mashable.com/article/smart-toys-security-privacy/

15 Shipley, "Reminder: Smart Toys are Cute, Cuddly, and Full of Security Risks"

16 "Consumer Notice: Internet-Connected Toys Could Present Privacy and Contact Concerns for Children," Public Service Announcement, Federal Bureau of Investigation, July 17, 2017, https:// www.ic3.gov/media/2017/170717.aspx

17 "Consumer Notice: Internet-Connected Toys Could Present Privacy and Contact Concerns for Children"

18 Sam Garin, "Stop the Expansion of Online Pre-K," Campaign for a Commercial-Free Childhood, June 25, 2019, https://commercialfreechildhood.org/stop-the-expansion-of-online-pre-k-2/

19 Rose Eveleth, "Facing Tomorrow's High-Tech Surveillance," *Vice*, October 29, 2018 https:// www.vice.com/en_us/article/j53ba3/facial-recognition-school-surveillance-v25n3

20 Eveleth, "Facing Tomorrow's High-Tech Surveillance"

21 "40 Organizations Release Principles for School Safety, Privacy, and Equity," The Future of Privacy Forum, March 27, 2019, https://fpf.org/2019/03/27/40-organizations-release-principles -for-school-safety-privacy-and-equity/

22 "40 Organizations Release Principles for School Safety, Privacy, and Equity"

23 Sara Collins and Amelia Vance, "40 Organizations Release Privacy Principles for Student Safety," Student Privacy Compass, March 26, 2019, http://studentprivacycompass.org/ schoolsafetyprinciples/

24 See, e.g., Melinda D. Anderson, "When School Feels Like Prison," *Atlantic*, September 12, 2016, https://www.theatlantic.com/education/archive/2016/09/when-school-feels-like-prison /499556/; and Tyler Kingkade, "The False Alarms That Get Kids Arrested,"

Atlantic, October 21, 2019, https://www.theatlantic.com/politics/archive/2019/10/
fake-school-shooting-threats-getting-kids-arrested/600238/

25 Louise Moon, "Pay Attention at the Back: Chinese School Installs Facial Recognition Cameras
to Keep an Eye on Pupils," *South China Morning Post*, May 16, 2018, https://www.scmp.com
/news/china/society/article/2146387/pay-attention-back-chinese-school-installs-facial-recognition

26 Moon, "Pay Attention at the Back: Chinese School Installs Facial Recognition Cameras to Keep
an Eye on Pupils"

27 Shawn De La Rosa, "Researchers Assessing AI's Ability to Measure Student Engagement," Education Dive, May 7, 2019, https://www.educationdive.com/news/researchers-assessing-ais-ability
-to-measure-student-engagement/554215/

28 Gianluca Mezzofiore, "Teachers Are Using Facial Recognition to See If Students Are Paying
Attention," *Mashable*, June 1, 2017, https://mashable.com/2017/06/01/facial-recognition-school
-france-ai-nestor/

29 Mezzofiore, "Teachers Are Using Facial Recognition to See If Students Are Paying Attention"

30 Mezzofiore, "Teachers Are Using Facial Recognition to See If Students Are Paying Attention"

31 Eveleth, "Facing Tomorrow's High-Tech Surveillance"

32 "Artificial Intelligence Market in the US Education Sector 2018–2022," Research and Markets,
August 2018, https://www.researchandmarkets.com/reports/4613290/artificial-intelligence
-market-in-the-us

33 Alyson Kline, "Can Artificial Intelligence Predict Student Engagement? Researchers Investigate," *Education Week*, May 6, 2019, https://blogs.edweek.org/edweek/DigitalEducation/2019/05
/artificial-intelligence-behavior-student-engagement.html?r=332999161

34 Benjamin Herold, "Schools Collect Tons of Student Information. Deleting It All Is a Major
Challenge," *Education Week*, March 15, 2019, https://blogs.edweek.org/edweek/DigitalEducation
/2019/03/schools_data_deletion.html?r=1646640126

35 Benjamin Herold, "Maryland Dad Wants June 30 to Be "National Student Data Deletion Day,"
Education Week, June 30, 2017, https://blogs.edweek.org/edweek/DigitalEducation/2017/06
/dad_wants_june_30_student_data_deletion_day.html?r=1156327556

CHAPTER 11

1 Drew Costley, "Smart Toilets Are Revealing the Health Data That Fitness Wearables Can't,"
One Zero, January 24, 2020, https://onezero.medium.com/smart-toilets-are-revealing-the-health
-data-that-wearables-cant-6113aa387323

2 Jonathan Rabinovitz, "Your Computer May Know You Have Parkinson's. Shall It Tell You?,"
Stanford Magazine, July 5, 2018, https://medium.com/stanford-magazine/your-computer-may
-know-you-have-parkinsons-shall-it-tell-you-e8f8907f4595

3 Alan Campbell, "Was Facial Recognition Being Used at Richmond Centre?" *Richmond News*,
September 1, 2018, https://www.richmond-news.com/news/was-facial-recognition-technology
-being-used-at-richmond-centre-1.23419076

4 Steve Knopper, "Why Taylor Swift Is Using Facial Recognition at Concerts," *Rolling Stone*, December 13, 2018, https://www.rollingstone.com/music/music-news/taylor-swift-facial -recognition-concerts-768741/

5 Knopper, "Why Taylor Swift Is Using Facial Recognition at Concerts"

6 TecSynt Solutions, "Pros and Cons of Facial Recognition Technology for Your Business," UpWork, December 27, 2017, https://www.upwork.com/hiring/for-clients/pros-cons-facial -recognition-technology-business/

7 TecSynt Solutions, "Pros and Cons of Facial Recognition Technology for Your Business"

8 Xuedong D. Huang, William H. Gates, Eric J. Horvitz, Joshua T. Goodman, Bradly A. Brunell, Susan T. Dumais, Gary W. Flake, Trenholme J. Griffin, and Oliver Hurst-Hiller, Web-Based Targeted Advertising in a Brick-and-Mortar Retail Establishment Using Online Customer Information, US Patent US20080004951A1, filed June 29, 2006, and issued October 24, 2006, https://patents.google.com/patent/US20080004951

9 Karen Gilchrist, "Alibaba Launches 'Smile to Pay' Facial Recognition System at KFC in China," CNBC, September 4, 2017, https://www.cnbc.com/2017/09/04/alibaba-launches-smile-to-pay -facial-recognition-system-at-kfc-china.html

10 "Privacy Is Under Threat from the Facial Recognition Revolution," *Financial Times*, October 3, 2017, https://www.ft.com/content/4707f246-a760-11e7-93c5-648314d2c72c

11 Shaun Walker, "Face Recognition App Taking Russia by Storm May Bring End to Public Ano-nymity," *Guardian*, May 17, 2016, https://www.theguardian.com/technology/2016/may/17 /findface-face-recognition-app-end-public-anonymity-vkontakte

12 Walker, "Face Recognition App Taking Russia by Storm May Bring End to Public Anonymity"

13 Patel v. Facebook, No. 18-15982, D.C. No. 3:15-cv-03747-JD, United States Court of Appeals for the Ninth Circuit, filed August 8, 2019, 17, http://cdn.ca9.uscourts.gov/datastore/opinions /2019/08/08/18-15982.pdf

14 Natasha Singer and Mike Isaac, "Facebook to Pay $550 Million to Settle Facial Recognition Lawsuit," *New York Times*, January 29, 2020, https://www.nytimes.com/2020/01/29/technology /facebook-privacy-lawsuit-earnings.html

15 See Jenni Ryall, "On No-Face Day, Employees Wear Masks to Avoid Fake Facial Expressions," *Mashable*, July 15, 2015, https://mashable.com/2015/07/15/no-face-day-china/; and Jeff John Rob-erts, "Our Facial Recognition Nightmare Is upon Us," *Fortune*, May 20, 2016, https://fortune .com/2016/05/20/facial-recognition-nightmare/

16 Antonio Regalado, "More Than 26 Million People Have Taken an At-Home Ancestry Test," *MIT Technology Review*, February 11, 2019, https://www.technologyreview.com/s/612880/more -than-26-million-people-have-taken-an-at-home-ancestry-test/

17 Regalado, "More Than 26 Million People Have Taken an At-Home Ancestry Test"

18 See, e.g., Sarah Zhang, "When White Nationalists Get DNA Tests That Reveal African Ances-try," *Atlantic*, August 17, 2017, https://www.theatlantic.com/science/archive/2017/08/white -nationalists-dna-ancestry/537108/; and Nsikan Akpan, "How White Supremacists Respond

When Their DNA Says They're Not 'White'," PBS, August 20, 2017, https://www.pbs.org /newshour/science/white-supremacists-respond-genetics-say-theyre-not-white

19 https://www.pooprints.com/

20 Ros Tamblyn, "DNA Tests Are Catching Dog Owners Who Don't Pick Up Poo," *BBC News*, May 2, 2019, https://www.bbc.com/news/av/stories-48122281/dna-tests-are-catching-dog-owners-who -don-t-pick-up-poo

21 See, e.g., Nicole Kobie, "The Complicated Truth About China's Social Credit System," *Wired*, June 7, 2019, https://www.wired.co.uk/article/china-social-credit-system-explained

22 Li Yuan, "Widespread Outcry in China over Death of Coronavirus Doctor," *New York Times*, February 7, 2020, https://www.nytimes.com/2020/02/07/business/china-coronavirus-doctor -death.html

23 "Wuhan Rounds Up the Infected as the Death Toll in China Jumps," *New York Times*, February 7, 2020, https://www.nytimes.com/2020/02/06/world/asia/coronavirus-china.html

24 See, e.g., Yingzhi Yang and Julie Zhu, "Coronavirus Brings China's Surveillance State out of the Shadows," *Reuters*, Febrary 7, 2020, https://www.reuters.com/article/us-china-health -surveillance/coronavirus-brings-chinas-surveillance-state-out-of-the-shadows -idUSKBN2011HO

25 Global Times, Twitter post, February 3, 2020, 9:43 p.m., https://twitter.com/globaltimesnews /status/1224569239253569538

26 Alison Rourke, "From Gassy Passengers to Viral Anthems, Beijing Seeks to Lighten the Mood Amid Crisis," *Guardian*, February 4, 2020, https://www.theguardian.com/world/2020/feb/04 /from-gassy-passengers-to-viral-anthems-beijing-seeks-to-lighten-mood-amid-crisis

27 Matt Novak, "That Video of Farts Captured by Thermal Cameras Looking for Coronavirus Is Totally Fake," *Gizmodo*, February 4, 2020, https://gizmodo.com/that-video-of-farts-captured -by-thermal-cameras-looking-1841442780

28 Aaron Mak, "All the Invasive Ways China Is Using Drones to Address the Coronavirus," *Slate*, February 4, 2020, https://slate.com/technology/2020/02/how-china-is-using-drones-to -contain-the-coronavirus.html

29 Mark Hanrahan, "Coronavirus: China Deploys Drones with Cameras, Loudhailers to Chastise People for Unsafe Behavior," *ABC News*, February 4, 2020, https://abcnews.go.com /International/coronavirus-china-deploys-drones-cameras-loudhailers-chastise-people/ story?id=68746989

30 "Watch: China Uses Talking Drones to Warn Citizens over Coronavirus," KDVR.com, February 1, 2020, https://kdvr.com/2020/02/01/watch-china-uses-talking-drones-to-warn-citizens -over-coronavirus/

31 Yang and Zhu, "Coronavirus Brings China's Surveillance State out of the Shadows"

32 Paul Mozur, Raymond Zhong, and Aaron Krolik, "In Coronavirus Fight, China Gives Citizens a Color Code with Red Flags," *New York Times*, March 1, 2020, https://www.nytimes .com/2020/03/01/business/china-coronavirus-surveillance.html

33 Yang and Zhu, "Coronavirus Brings China's Surveillance State out of the Shadows"

34 "Wuhan Rounds Up the Infected as the Death Toll in China Jumps"

35 "Half the Planet Is on Lockdown, But Not Every U.S. State Is, Even After Alabama Issues an Order," *New York Times*, April 3, 2020, https://www.nytimes.com/2020/04/03/world/coronavirus-news-updates.html

36 Sue Halpern, "Can We Track COVID-19 and Protect Privacy at the Same Time?" *New Yorker*, April 27, 2020, https://www.newyorker.com/tech/annals-of-technology/can-we-track-covid-19-and-protect-privacy-at-the-same-time

37 "Drones Used in Effort to Slow the Spread of COVID-19," *CBS News,* April 27, 2020, https://www.cbsnews.com/news/coronavirus-drones-slow-spread-covid-19/; Peter Lane Taylor, "Could 'Pandemic Drones' Help Slow Coronavirus? Probably Not – But COVID-19 IS a Boom for Business," *Forbes*, April 25, 2020, https://www.forbes.com/sites/petertaylor/2020/04/25/could-pandemic-drones-help-slow-coronavirus-probably-not-but-covid-19-is-a-boom-for-business/#7fc77edb62a4

38 Halpern, "Can We Track COVID-19 and Protect Privacy at the Same Time?"

39 "FDA's Digital Health Policies Allow Innovators to Create COVID-19 Related Public Health Solutions," Food and Drug Administration, March 26, 2020, https://www.fda.gov/medical-devices/digital-health/digital-health-policies-and-public-health-solutions-covid-19

40 Sara Burnett, "Michigan Militia Puts Armed Protest in the Spotlight," *Washington Post*, May 2, 2020, https://www.washingtonpost.com/politics/michigan-militia-puts-armed-protest-in-the-spotlight/2020/05/02/aa8fa700-8c32-11ea-80df-d24b35a568ae_story.html

CHAPTER 12

1 The United Kingdom now has a law, informally known as the Turing Act, that was passed in 2017 in order to retroactively pardon men like Turing who had suffered criminal convictions relating to their sexual orientation during the decades when homosexuality was outlawed.

2 Alan Turing, "Computing Machinery and Intelligence," *Mind* magazine, October 1950, https://www.turing.org.uk/scrapbook/test.html and https://www.abelard.org/turpap/turpap.php

3 Turing, "Computing Machinery and Intelligence"

4 James O'Malley, "Captcha If You Can: How You've Been Training AI for Years Without Realising It," *TechRadar*, January 12, 2018, https://www.techradar.com/news/captcha-if-you-can-how-youve-been-training-ai-for-years-without-realising-it

5 "History of Sweatshops, 1820-1880," National Museum of American History, Smithsonian, http://americanhistory.si.edu/sweatshops/history-1820-1880 (accessed April 17, 2020)

6 By the twenty-first century, the gender pay gap had only narrowed slightly, to an average of 20 percent. See, e.g., Ariane Hegewisch and Heidi Hartmann, "The Gender Wage Gap: 2018 Earnings Differences by Race and Ethnicity," Institute for Women's Policy Research, March 7, 2019, https://iwpr.org/publications/gender-wage-gap-2018/

7 "History of Sweatshops, 1820-1880"

8   https://www.surveyjunkie.com/

9   Jim Wang, "A Detailed Review of My Survey Junkie Experience (2020)," *Best Wallet Hacks*, April 17, 2020, https://wallethacks.com/survey-junkie-review/

10 Jeff Rose, "Survey Junkie Review," *Good Financial Cents*, April 10, 2020, https://www .goodfinancialcents.com/survey-junkie-review

11 "The Belmont Report," US Department of Health & Human Services, https://www.hhs.gov /ohrp/regulations-and-policy/belmont-report/index.html (accessed April 17, 2020)

12 "Frequently Asked Questions (FAQs)," NIH Grants & Funding, https://grants.nih.gov/faqs#/ (accessed April 17, 2020)

13 45 CFR § 46.116 - General requirements for informed consent, https://www.law.cornell.edu/cfr /text/45/46.116 (accessed April 17, 2020)

14 45 CFR § 46.111 - Criteria for IRB approval of research, Legal Information Institute, https:// www.law.cornell.edu/cfr/text/45/46.111 (accessed April 17, 2020)

CHAPTER 13

1   "Online and Digital Abuse," Women's Aid, https://www.womensaid.org.uk/information -support/what-is-domestic-abuse/onlinesafety/ (accessed April 17, 2020)

2   Aarti Shahani, "Smartphones Are Used to Stalk and Control Domestic Abuse Victims," NPR, September 15, 2014, https://www.npr.org/sections/alltechconsidered/2014/09/15/346149979 /smartphones-are-used-to-stalk-control-domestic-abuse-victims

3   Shahani, "Smartphones Are Used to Stalk and Control Domestic Abuse Victims"

4   Christopher Parsons, Adam Molnar, Jakub Dalek, Jeffrey Knockel, Miles Kenyon, Bennett Haselton, Cynthia Khoo, and Ron Deibert, "The Predator in Your Pocket," Citizen Lab, June 12, 2019, https://citizenlab.ca/2019/06/the-predator-in-your-pocket-a-multidisciplinary -assessment-of-the-stalkerware-application-industry/

5   Parsons et al., "The Predator in Your Pocket"

6   Joseph Cox, "Twitter Pushed Adverts for Spyware to Monitor Girlfriends," *Vice*, July 3, 2019, https://www.vice.com/en_us/article/3k3wx5/twitter-pushed-adverts-for-spyware-to-track -girlfriends

7   Parsons et al., "The Predator in Your Pocket"

8   Parsons et al., "The Predator in Your Pocket," 15

9   Jacqueline Connor, "Who's Stalking: What to Know About Mobile Spyware," Federal Trade Commission Consumer Information, September 26, 2016, https://www.consumer.ftc.gov /blog/2016/09/whos-stalking-what-know-about-mobile-spyware

10 See, e.g., Marlisse Silver Sweeney, "What the Law Can and Can't Do About Online Harass- ment," *Atlantic,* November 12, 2014, https://www.theatlantic.com/technology/archive/2014/11 /what-the-law-can-and-cant-do-about-online-harassment/382638/

11 Andy Greenberg, "The Simple Way Apple and Google Let Domestic Abusers Stalk Victims," *Wired*, July 2, 2019, https://www.wired.com/story/common-apps-domestic-abusers-stalk-victims/

12 See, e.g., Hala Aldosari, "Guardians of the Gender Gap," *Foreign Affairs*, August 10, 2016, https://www.foreignaffairs.com/articles/saudi-arabia/2016-08-10/guardians-gender-gap; and "End Male Guardianship," Human Rights Watch, https://www.hrw.org/endmaleguardianship (accessed April 17, 2020)

13 "See, e.g., Emma Grey Ellis, "It's Time for Facebook to Deal with the Grimy History of Revenge," *Wired*, March 14, 2017, https://www.wired.com/2017/03/revenge-porn-facebook/; Alexis Tsoulis-Ray, "A Brief History of Revenge Porn," *New York Magazine*, July 19, 2013, https://nymag.com/news/features/sex/revenge-porn-2013-7/; and Heather Kuldell, "Feds Target Revenge Porn Website," NextGov.com, January 9, 2018. https://www.nextgov.com/policy/2018/01/feds-target-revenge-porn-website/145080/

14 See, e.g., Mudasir Kamal and William J. Newman, "Revenge Pornography: Mental Health Implications and Related Legislation," *Journal of the American Academy of Psychiatry and the Law*, September 2016, http://jaapl.org/content/44/3/359; and Danielle Keats Citron and Mary Anne Franks, "Criminalizing Revenge Porn," *49 Wake Forest Law Review 345* (2014), https://digitalcommons.law.umaryland.edu/fac_pubs/1420/

15 Jeff Kosseff, *Twenty-Six Words That Created the Internet* (New York: Cornell University Press, 2019)

16 Jeff Kosseff, "What's in a Name? Quite a Bit, If You're Talking About Section 230," *Lawfare*, December 19, 2019, https://www.lawfareblog.com/whats-name-quite-bit-if-youre-talking-about-section-230

17 Andy Greenberg, "It's Been 20 Years Since This Man Declared Cyberspace Independence," *Wired*, February 8, 2016, https://www.wired.com/2016/02/its-been-20-years-since-this-man-declared-cyberspace-independence/

18 John Perry Barlow, "A Declaration of the Independence of Cyberspace," *Electronic Frontier Foundation*, February 8, 1996, https://www.eff.org/cyberspace-independence

19 Michael Brice-Sadler, Avi Selk, and Eli Rosenberg, "Prankster Sentenced to 20 Years for a Fake 911 Call That Led Police to Kill an Innocent Man," *Washington Post*, March 29, 2019, www.washingtonpost.com/nation/2019/03/29prankster-sentenced-years-fake-call-that-led-police-kill-an-innocent-man/

20 Nellie Bowles, "How 'Doxxing' Became a Mainstream Tool in the Culture Wars," *New York Times*, August 30, 2017, https://www.nytimes.com/2017/08/30/technology/doxxing-protests.html

21 Kamal and Newman, "Revenge Pornography: Mental Health Implications and Related Legislation"

22 Barlow, "Declaration of the Independence of Cyberspace"

23 Greenberg, "It's Been 20 Years Since This Man Declared Cyberspace Independence"

24 Zoë Quinn, "Zoë Quinn: What Happened After GamerGate Hacked Me," *Time*, September 11, 2017, https://time.com/4927076/zoe-quinn-gamergate-doxxing-crash-override-excerpt/

25 Elise Viebeck, "Kavanaugh's Accuser Thought Her Life Would be Upended. She Was Right," *Washington Post*, September 28, 2018, https://www.washingtonpost.com/politics

/kavanaughs-accuser-thought-her-life-would-be-upended-she-was-right/2018/09/18
/1f0a824e-bb5b-11e8-a8aa-860695e7f3fc_story.html

26 "The Global Gender Gap Report 2018," World Economic Forum, http://www3.weforum.org
/docs/WEF_GGGR_2018.pdf (accessed April 17, 2020)

27 "Saudi Women Get ID Cards," *BBC News*, December 10, 2001, http://news.bbc.co.uk/2/hi
/middle_east/1702342.stm

28 Jassim Abuzaid, "IDs a Must for Saudi Women," *Arab News*, March 26, 2013, http://www
.arabnews.com/news/446108

29 "Saudi Arabia Gives Women the Right to a Copy of Their Marriage Contract," *Guardian*, May
3, 2016, https://www.theguardian.com/world/2016/may/03
/saudi-arabia-gives-women-the-right-to-a-copy-of-their-marriage-contract

30 See, e.g., "Freedom on the Net 2018 – Saudi Arabia," Freedom House, https://freedomhouse.org
/country/saudi-arabia/freedom-net/2018 (accessed April 17, 2020)

31 "Saudi Arabia: Mobile App Keeps Saudi Women at Home," Human Rights Watch, May 6, 2019,
https://www.hrw.org/news/2019/05/06/saudi-arabia-mobile-app-keeps-women-home#

CHAPTER 14

1   "Records, Computers, and the Rights of Citizens"

2   "Records, Computers, and the Rights of Citizens"

3   Both terms—FIPPs and FIPs—are in widespread use. For example, the US Federal Trade Commission refers to the FIPPs, while organizations such as the World Privacy Forum, a privacy advocacy group, refers to the FIPs.

4   "Records, Computers, and the Rights of Citizens"

5   Seymour M. Hersh, "Huge C.I.A. Operation Reported in U.S. Against Antiwar Forces, Other Dissidents in Nixon Years," *New York Times*, December 22, 1974, https://www.nytimes
.com/1974/12/22/archives/huge-cia-operation-reported-in-u-s-against-antiwar-forces-other.html

6   Hersh, "Huge C.I.A. Operation Reported in U.S. Against Antiwar Forces, Other Dissidents in Nixon Years"

7   Hersh, "Huge C.I.A. Operation Reported in U.S. Against Antiwar Forces, Other Dissidents in Nixon Years"

8   U.S. Senate Subcommittee on Intelligence and the Rights of Americans of the Select Committee on Intelligence (Church Committee), "Huston Plan: Church Committee Hearings, Vol. 2," Internet Archive, September 1975, https://archive.org/details/Church-Committee-Hearings
-Volume2-Huston-Plan/page/n5

9   Huston Plan, Hearings Before the Select Committee to Study Governmental Operations with Respect to Intelligence Activities of the United States Senate Ninety-Fourth Congress First Session, Volume 2, Intelligence Activities, Senate Resolution 21, September 23–25, 1975, 141–183, https://www.intelligence.senate.gov/sites/default/files/94intelligence_activities_II.pdf

10 Huston Plan, 159

11 Huston Plan, 155–156

12 Huston Plan, 194

13 Huston Plan, 194

14 Huston Plan, 195–196

15 Huston Plan, 193

16 In a particularly tantalizing detail, the memo's recommendation on break-ins noted that "The FBI, in Mr. Hoover's younger days, used to conduct such operations with great success and with no exposure. The information secured was invaluable." Huston Plan, 195

17 Huston Plan, 198–200

18 Church Committee, "Huston Plan: Church Committee Hearings, Vol. 2"

19 Church Committee, Senate Resolution 21, January 27, 1975, https://www.senate.gov /artandhistory/history/common/investigations/pdf/ChurchCommittee_SRes21.pdf

20 Huston Plan, 245–253

21 See, generally, Huston Plan

22 The Pike Committee carried out similar work but never issued a final report. Copies of the Pike Committee's draft report made their way into the hands of various news outlets and are widely available on the internet. See, e.g., House Select Committee on Intelligence, "Pike Committee Report Full," Internet Archive, 1976, https://archive.org/details/PikeCommitteeReportFull/

23 "Final Report of the Select Committee to Study Governmental Operations with Respect to Intelligence Activities, United States Senate, Together with Additional, Supplemental, and Separate Views," Foreign and Military Intelligence, Book I, April 26, 1976, 1 https://www .intelligence.senate.gov/sites/default/files/94755_I.pdf

24 "Final Report of the Select Committee to Study Governmental Operations," Book I, 1

25 "Final Report of the Select Committee to Study Governmental Operations," Book I, 2

26 "Final Report of the Select Committee to Study Governmental Operations with Respect to Intelligence Activities, United States Senate, Together with Additional, Supplemental, and Separate Views," Intelligence Activities and the Rights of Americans, Book II, April 26, 1976, iii, https://www.intelligence.senate.gov/sites/default/files/94755_II.pdf

27 "Final Report of the Select Committee to Study Governmental Operations," Book II, 1

28 "Final Report of the Select Committee to Study Governmental Operations," Book II, 2–3

29 "Final Report of the Select Committee to Study Governmental Operations," Book II, 3–4

30 "Final Report of the Select Committee to Study Governmental Operations," Book II, 5–6

31 "Final Report of the Select Committee to Study Governmental Operations," Book II, 5–6

32 "Final Report of the Select Committee to Study Governmental Operations," Book II, 6–7

33 "Final Report of the Select Committee to Study Governmental Operations," Book II, 7–9

34 "Final Report of the Select Committee to Study Governmental Operations," Book II, 9–10

35 "Final Report of the Select Committee to Study Governmental Operations," Book II, 16–20

36 "Final Report of the Select Committee to Study Governmental Operations," Book II, 20

37 The committee's report also described a number of other techniques, such as secret informants, infiltrating groups, and covert action, which are less directly related to data privacy, and therefore are outside the scope of this book.

38 "Pike Committee Report Full," 165

39 "Pike Committee Report Full," 181

CHAPTER 15

1   "Joint Inquiry into Intelligence Community Activities Before and After the Terrorist Attacks
    of September 11, 2001," Report of the U.S. Senate Select Committee on Intelligence And U.S.
    House Permanent Select Committee on Intelligence Together with Additional Views, December
    2002, xvi–xvii, https://fas.org/irp/congress/2002_rpt/911rept.pdf

2   "Joint Inquiry into Intelligence Community Activities Before and After the Terrorist Attacks of
    September 11, 2001," 36

3   "Joint Inquiry into Intelligence Community Activities Before and After the Terrorist Attacks of
    September 11, 2001"

CHAPTER 16

1   Carpenter v. United States, 585 U.S. __(2018), https://www.supremecourt.gov/opinions/17pdf
    /16-402_h315.pdf

2   James Risen and Eric Lichtblau, "Bush Lets U.S. Spy on Callers Without Courts," *New York
    Times*, December 16, 2005, https://www.nytimes.com/2005/12/16/politics/bush-lets-us-spy-on
    -callers-without-courts.html

3   What the Constitution requires is that searches and seizures be reasonable, and a particularized
    warrant is one way in which reasonableness can be demonstrated. But American law has a long
    history of recognizing other valid ways to carry out a search: drunk driving checkpoints, school
    locker searches, searches incident to arrest—all of these and more presented examples of a long
    line of jurisprudence explaining the reasons why the government doesn't always require a war-
    rant in order to carry out a search.

4   Senate Select Committee on Intelligence, Report to accompany S. 2248, the Foreign Intelligence
    Surveillance Act of 1978, Amendments Act of 2007, 4, https://www.congress.gov/110/crpt
    /srpt209/CRPT-110srpt209.pdf

5   Senate Select Committee on Intelligence, Report to accompany S. 2248, 5

6   Senate Select Committee on Intelligence, Report to accompany S. 2248, 7–8

7   Statement for the Record of J. Michael McConnell, Director of National Intelligence, House
    Judiciary Committee Hearing on the Foreign Intelligence Surveillance Act and Protect America
    Act, September 18, 2007, 4–5, https://www.dni.gov/files/documents/Newsroom/Testimonies
    /20070918_testimony.pdf

8   Statement for the Record of J. Michael McConnell, 4–5

9   "Report on the Surveillance Program Operated Pursuant to Section 702 of the Foreign Intelli-
    gence Surveillance Act," Privacy and Civil Liberties Oversight Board, July 2, 2014, 25, https://
    www.pclob.gov/library/702-Report-2.pdf

10 Statement of April F. Doss, Partner, Saul Ewing, LLP, Before the United States House of Representatives Judiciary Committee, Concerning Section 702 of the Foreign Intelligence Surveillance Act, March 1, 2017, https://republicans-judiciary.house.gov/wp-content/uploads /2017/02/Doss-Testimony.pdf

11 The US person may only be intentionally targeted with proper authorization, which in nearly all cases requires a FISC order or similar court approval (such as a criminal wiretap authorization under Title III of the Wiretap Act). Some critics have raised concerns about the possibility of "reverse targeting"—that is, targeting a foreign person or entity in order to obtain their communications with a U.S. person. Reverse targeting is, and always has been, unlawful under FISA, as well as prohibited under the procedures for targeting under EO 12333. In addition to being barred, it is precisely the type of activity that overseers look for, both in the internal agency compliance organizations as well as external overseers. The chances that a rogue analyst, or any entire agency, could get away with reverse targeting undetected are, thanks to the robust oversight mechanisms in place, extremely slim.

12 Permanent Select Committee on Intelligence, Report 95–1283, Pt. 1, the Foreign Intelligence Surveillance Act of 1978, June 8, 1978, 19–20, https://fas.org/irp/agency/doj/fisa/hspci1978.pdf

13 Permanent Select Committee on Intelligence, Report 95–1283, Pt. 1, 58

14 Statement of April F. Doss

CHAPTER 17

1 Tom Jackman, "Police Use of 'Stingray' Cellphone Tracker Requires Search Warrant Appeals Court Rules," *Washington Post*, September 21, 2017, https://www.washingtonpost.com/news /true-crime/wp/2017/09/21/police-use-of-stingray-cellphone-tracker-requires-search-warrant -appeals-court-rules/

2 Rick Rojas, "In Newark, Police Cameras, and the Internet, Watch You," *New York Times*, June 9, 2018, https://www.nytimes.com/2018/06/09/nyregion/newark-surveillance-cameras-police .html

3 See, e.g., Amanda Riley, "A Big Test of Police Body Cameras Defies Expectations," *New York Times*, October 20, 2017, https://www.nytimes.com/2017/10/20/upshot/a-big-test-of-police-body -cameras-defies-expectations.html

4 See, e.g., Nicole Raz, "Las Vegas Police Drones Will Monitor New Year's Eve Crowds," *Las Vegas Review Journal*, December 27, 2017, https://www.reviewjournal.com/entertainment/new-years -eve-in-vegas/las-vegas-police-drones-will-monitor-new-years-eve-crowds/

5 Jake Laperruque and David Janovsky, "These Police Drones Are Watching You," *Pogo*, September 25, 2018, https://www.pogo.org/analysis/2018/09/these-police-drones-are-watching-you/

6 In addition to privacy concerns, police use of drones raises other potential issues as well, ranging from the need to address what could become congested airspace, to questions about liability for damages caused by the use of drones, to the potential national security risks associated with the fact that a number of drones are manufactured by adversarial foreign nations.

7   See, e.g., Jean Marie Laskas, "Inside the Federal Bureau of Way Too Many Guns," *GQ*, August 30, 2016, https://www.gq.com/story/inside-federal-bureau-of-way-too-many-guns

8   Drew Harwell, "Doorbell-Camera Firm Ring Has Partnered with 400 Police Forces, Extending Surveillance Concerns," *Washington Post*, August 28, 2019, https://www.washingtonpost.com/technology/2019/08/28/doorbell-camera-firm-ring-has-partnered-with-police-forces-extending-surveillance-reach/

9   Catherine Thorbecke, "Senator Blasts Amazon's Ring Doorbell as Open Door for Privacy and Civil Liberties Violations," *ABC News*, November 20, 2019, https://abcnews.go.com/Business/senator-blasts-amazons-ring-doorbell-open-door-privacy/story?id=67162384

10 Neil Vigdor, "Somebody's Watching: Hackers Breach Home Security Cameras," *New York Times*, December 15, 2019, https://www.nytimes.com/2019/12/15/us/Hacked-ring-home-security-cameras.html

11 Richard Kenny, "Residents Snapping Every Passing Car—Security or Spying?" *BBC News*, July 7, 2019, https://www.bbc.com/news/av/stories-48885776/residents-snapping-every-passing-car-security-or-spying

12 Harris v. U.S., 390 U.S. 234 (1968) (plain view); and Oliver v. U.S., 466 U.S. 170 (1984) (open fields)

13 Smith v. Maryland, 442 U.S. 735 (1979)

14 California v. Ciraolo, 476 U.S. 207 (1986); Florida v. Riley, 488 U.S. 445 (1989); and Dow Chemical v. U.S., 476 U.S. 227 (1986)

15 Kyllo v. U.S., 533 U.S. 27 (2001)

16 Kyllo v. U.S., 533 U.S. 27 (2001)

17 United States v. Jones, 565 U.S. 400 (2012)

18 See, e.g., Olmstead v. U.S., 277 U.S. 438 (1928)

19 See, e.g., Katz v. U.S., 389 U.S. 347 (1967)

20 Trisha Thadani, "San Francisco Bans City Use of Facial Recognition Surveillance Technology," *San Francisco Chronicle*, May 14, 2019, https://www.sfchronicle.com/politics/article/San-Francisco-bans-city-use-of-facial-recognition-13845370.php

CHAPTER 18

1   Jones v. U.S., 565 U.S. 400 (2012)

2   Riley v. California, 573 U.S. 373 (2014)

3   Yves-Alexandre de Montjoye, César A. Hidalgo, Michel Verleysen, and Vincent D. Blondel, "Unique in the Crowd: The Privacy Bounds of Human Mobility," *Nature*, March 25, 2013, https://www.nature.com/articles/srep01376

4   Caitlin Oprysko, Anita Kumar, and Nahal Toosi, "Trump Administration Expands Travel Ban," *Politico*, January 31, 2020, https://www.politico.com/news/2020/01/31/trump-administration-expands-travel-ban-110005

5   Nick Miroff, "Trump Administration Plans to Announce 100-Mile Mark on Border Wall Construction," *Washington Post*, January 9, 2020, https://www.washingtonpost.com

/immigration/trump-administration-plans-to-announce-100-mile-mark-on-border-wall
-construction/2020/01/09/9ed58874-3301-11ea-898f-eb846b7e9feb_story.html

6   Ron Nixon, "Cellphone and Computer Searches at US Border Rise Under Trump," *New York Times*, January 5, 2018, https://www.nytimes.com/2018/01/05/us/politics/trump-border
-search-cellphone-computer.html

7   Riley v. California, 573 U.S. 373 (2014), 8–9, https://www.supremecourt.gov/opinions
/13pdf/13-132_8l9c.pdf

8   Riley v. California, 17

9   Riley v. California, 18

10   Riley v. California, 20

11   Riley v. California, 28

12   See, e.g., Boyd v. U.S., 116 U.S. 616 (1886); U.S. v. Ramsey, 431 U.S. 606 (1977); and U.S. v. Montoya de Hernandez, 473 U.S. 531 (1985)

13   Hearing Transcript, Senate Homeland Security and Government Affairs Committee Hearing on Homeland Security and Public Safety, April 5, 2017, 14, https://www.thisweekinimmigration
.com/uploads/6/9/2/2/69228175/hearingtranscript_senatehomelandsecurityandgovernmental
affairsheaingwithsecretarykelly_2017-04-05.pdf

14   Tanvi Misra, "Inside the Massive US 'Border Zone'," *City Lab*, May 14, 2018, https://www
.citylab.com/equity/2018/05/who-lives-in-border-patrols-100-mile-zone-probably-you-mapped
/558275/

15   See, e.g., Eric Westervelt, "As Migrants Stream In at the Border, Inland Checkpoints Feel the Strain," NPR, June 12, 2019, https://www.npr.org/2019/06/12/731797754/as-migrants-stream-in
-at-the-border-inland-checkpoints-feel-the-strain

16   Alasaad v. McAlaneen, United States District Court, District of Massachusetts, No. 17-cv-11730-
DJC, November 12, 2019, 25, https://www.aclu.org/legal-document/alasaad-v-mcaleenan
-opinion-summary-judgment

17   Alasaad v. McAlaneen, 29–30

18   Alasaad v. McAlaneen, 29–32

19   Donald J. Trump, Twitter post, July 14, 2019, 5:27 a.m., https://twitter.com/realDonaldTrump/
status/1150381395078000643

20   Megan Molteni, "How DNA Testing at the US-Mexico Border Will Actually Work," *Wired*, May 2, 2019, https://www.wired.com/story/how-dna-testing-at-the-us-mexico-border-will
-actually-work/

21   Molteni, "How DNA Testing at the US-Mexico Border Will Actually Work"

22   Nila Milanich, "'Rapid DNA' Promises to Identify Fake Families at the Border. It Won't," *Washington Post*, May 13, 2019, https://www.washingtonpost.com/outlook/2019/05/13/rapid
-dna-promises-identify-fake-families-border-it-wont/

23   See, e.g., Patrick J. Lyons, "Trump Wants to Abolish Birthright Citizenship. Can He Do That?" *New York Times*, August 22, 2019, https://www.nytimes.com/2019/08/22/us/birthright
-citizenship-14th-amendment-trump.html; and Erin Blakemore, "Why the United States Has

Birthright Citizenship," *History*, October 30, 2018, https://www.history.com/news/birthright
-citizenship-history-united-states

24 Caitlin Dickerson, "US Government Plans to Collect DNA from Detained Immigrants," *New
York Times*, October 2, 2019, https://www.nytimes.com/2019/10/02/us/dna-testing-immigrants
.html

25 Richard Perez-Pena, "Contrary to Trump's Claims, Immigrants Are Less Likely to Commit
Crimes," *New York Times*, January 26, 2017, https://www.nytimes.com/2017/01/26/us/trump
-illegal-immigrants-crime.html

CHAPTER 19

1   "The Reich SS Leader," Topographie des Terrors, https://www.topographie.de/en/the
-historic-site/reich-ss-leader/ (accessed April 17, 2020)

2   "The Reich SS Leader"

3   "The Secret State Police (Gestapo)," Topographie des Terrors, https://www.topographie.de/en
/the-historic-site/gestapo/ (accessed April 17, 2020)

4   "Art. 9 GDPR – Processing of Special Categories of Personal Data," Intersoft Consulting,
https://gdpr-info.eu/art-9-gdpr/ (accessed April 17, 2020)

5   "Press Release No 70/14," Court of Justice of the European Union, May 13, 2014, https://curia
.europa.eu/jcms/upload/docs/application/pdf/2014-05/cp140070en.pdf

6   "Press Release No 70/14"

7   "Press Release No 70/14"

8   Daniel Boffey, "Dutch Surgeon Wins Landmark 'Right to Be Forgotten' Case," *Guardian*, January 21, 2019, https://www.theguardian.com/technology/2019/jan/21/dutch-surgeon-wins
-landmark-right-to-be-forgotten-case-google

9   Sarah Marsh, "'Right to Be Forgotten' on Google Only Applies in EU, Court Rules," *Guardian*,
September 24, 2011, https://www.theguardian.com/technology/2019/sep/24
/victory-for-google-in-landmark-right-to-be-forgotten-case

10 http://www.oecdprivacy.org/

11 See, e.g., Jean-Paul Fitoussi and Edmund Strother Phelps, "Causes of the 1980s Slump in
Europe," *Brookings Papers on Economic Activity* 2 (1986), https://www.brookings.edu/wp
-content/uploads/1986/06/1986b_bpea_fitoussi_phelps_sachs.pdf

12 John R. Schindler, "Social Media Is Helping Putin Kill Our Democracy," *Observer*, February 6,
2018, https://observer.com/2018/02/robert-hannigan-explains-how-russia-vladimir-putin
-uses-social-media/

13 "Recital 2—Respect of the Fundamental Rights and Freedoms," Intersoft Consulting, https://
gdpr-info.eu/recitals/no-2/ (accessed April 17, 2020)

14 The language from the Regulation states that its goals are the fostering of "economic and social
progress, [and] the strengthening and the convergence of the economies with the internal [Euro-
pean] market." "Recital 2—Respect of the Fundamental Rights and Freedoms"

15 For example, the Poland data privacy regulator levied a steep fine against a company for collect-ing information that was already publicly posted on websites of other government entities. See Natasha Lomas, "Covert Data-Scraping on Watch as EU DPA Lays Down 'Radical' GDPR Red Line," TechCrunch, March 30, 2019, https://techcrunch.com/2019/03/30/covert-data-scraping-on -watch-as-eu-dpa-lays-down-radical-gdpr-red-line/

16 In the *Schrems II* litigation, the Advocate General wrote that access to a communications stream for packet-level filtering purposes "constitutes in my view an interference" with the right to privacy, even when the packets were only inspected quickly via automated means and the infor-mation wasn't retained. Data Protection Commissioner v. Facebook Ireland, Case C-311/18, Opinion of Advocate General, paragraph 260, Dec. 19, 2019, http://curia.europa.eu/juris /document/document.jsf?text=&docid=221826&pageIndex=0&doclang=EN&mode =req&dir=&occ=first&part=1&cid=49246#Footnote129

CHAPTER 20

1 See, e.g., Adrian Shahbaz, "The Rise of Digital Authoritarianism," Freedom House, https:// freedomhouse.org/report/freedom-net/2018/rise-digital-authoritarianism (accessed April 17, 2020)

2 "Privatizing Censorship, Eroding Privacy," Freedom House, https://freedomhouse.org/report /freedom-net/2015/privatizing-censorship-eroding-privacy (accessed April 17, 2020)

3 "Freedom on the Net 2018 – Bahrain," Freedom House, https://freedomhouse.org/country/ bahrain/freedom-net/2018 (accessed April 17, 2020)

4 "Freedom on the Net 2018 – Ethiopia," Freedom House, https://freedomhouse.org/country /ethiopia/freedom-net/2018 (accessed April 17, 2020)

5 "Freedom on the Net 2018 – Cuba," Freedom House, https://freedomhouse.org/country/cuba /freedom-net/2018 (accessed April 17, 2020)

6 "Freedom on the Net 2018 – Vietnam," Freedom House, https://freedomhouse.org/country /vietnam/freedom-net/2018 (accessed April 17, 2020)

7 "Freedom on the Net 2018 – United Arab Emirates," Freedom House, https://freedomhouse.org /country/united-arab-emirates/freedom-net/2018 (accessed April 17, 2020)

8 See, e.g., Keith Breene, "Who Are the Cyberwar SuperPowers?" World Economic Forum, May 4, 2016, https://www.weforum.org/agenda/2016/05/who-are-the-cyberwar-superpowers/

9 James R. Clapper, Marcel Lettre, and Michael S. Rogers, "Joint Statement for the Record to the Senate Armed Services Committee, Foreign Cyber Threats to the United States," United States Senate Committee on Armed Services, January 5, 2017, https://www.armed-services.senate.gov /imo/media/doc/Clapper-Lettre-Rogers_01-05-16.pdf

10 "2018 Global Cybersecurity Index (v3)," International Telecommunications Union, https://www .itu.int/en/ITU-D/Cybersecurity/Pages/global-cybersecurity-index.aspx (accessed April 17, 2020)

11 Clapper, Lettre, and Rogers, "Joint Statement"

12　There's no question that all Western liberal democracies fall short of those ideals. However, they all proclaim them and, to varying degrees under different government leadership, aspire to them.

13　Nathalie Marechal, "Are You Upset About Russia Interfering in Our Elections?" *Slate*, March 20, 2017, https://slate.com/technology/2017/03/russias-election-interfering-cant-be-separated -from-its-domestic-surveillance.html

14　Andrei Soldatov and Irina Borogan, "Russia's Surveillance State," *World Policy*, September 12, 2013, https://worldpolicy.org/2013/09/12/russias-surveillance-state/

15　Soldatov and Borogan, "Russia's Surveillance State"

16　Neil McFarquhar, "Russian Quietly Tightens the Reins on the Web with Bloggers Law," *New York Times*, May 6, 2014, https://www.nytimes.com/2014/05/07/world/europe/russia-quietly -tightens-reins-on-web-with-bloggers-law.html

17　See, e.g., Maria Repnikova, "How Chinese Authorities and Individuals Use the Internet," Hoover Institution, October 29, 2018. https://www.hoover.org/research/how-chinese-authorities -and-individuals-use-internet

18　See, e.g., Daniel Wagner, "What China's Cybersecurity Laws Say About the Future," International Policy Digest, May 13, 2019, https://intpolicydigest.org/2019/05/13/what-china-s -cybersecurity-law-says-about-the-future/

19　See, e.g., Amy Gunia, "The Tiananmen Massacre Is One of China's Most Censored Topics. Here's a Look at What Gets Banned," *Time*, April 17, 2019 https://time.com/5571372/tiananmen -massacre-june-4-1989-china-censorship/

20　Elaine Yu, "Surveillance-Savvy Protesters Go Digitally Dark," *Yahoo! News*, June 12, 2019, https://news.yahoo.com/surveillance-savvy-hong-kong-protesters-digitally-dark-003014805.html

21　Justin Sherman and Robert Morgus, "Authoritarians Are Exporting Surveillance Tech, and with It Their Vision for the Internet," Council on Foreign Relations, December 5, 2018, https://www .cfr.org/blog/authoritarians-are-exporting-surveillance-tech-and-it-their-vision-internet

22　https://www.wassenaar.org/

CHAPTER 21

1　See, e.g., Natasha Lomas, "Covert Data-Scraping on Watch as EU DPA Lays Down 'Radical' GDPR Red-Line," TechCrunch, March 30, 2019, https://techcrunch.com/2019/03/30/covert -data-scraping-on-watch-as-eu-dpa-lays-down-radical-gdpr-red-line/

2　Spokeo, Inc. v. Robbins, 578 US ___ (2016), 136 S. Ct. 1540, 1549, https://www.supremecourt .gov/opinions/15pdf/13-1339dif_3m92.pdf

3　Patel v. Facebook, No. 18-15982

4　In re. Facebook, Inc., Consumer Privacy User Profile Litigation

5　Jane Yakowitz Bambauer, "The New Intrusion," *Notre Dame Law Review* 88, no. 1 (2013), 209, http://ndlawreview.org/wp-content/uploads/2013/07/NDL105.pdf

6　Rachel Levinson-Waldman, "Hiding in Plain Sight: A Fourth Amendment Framework for Analyzing Government Surveillance in Public," *Emory Law Journal* 66, no. 527, http://law.emory.edu /elj/_documents/volumes/66/3/levinson-waldman.pdf

7 McCann and Hall, "Blocking the Data Stalkers"

8 McCann and Hall, "Blocking the Data Stalkers," 2

9 J. Sterling Livingston, "Pygmalion in Management," *Harvard Business Review*, January 2003, https://hbr.org/2003/01/pygmalion-in-management

10 See, e.g., "Report on the Telephone Records Program Conducted Under Section 215 of the USA PATRIOT Act and on the Operations of the Foreign Intelligence Surveillance Court," Privacy and Civil Liberties Oversight Board, January 23, 2014, 11, https://www.pclob.gov/library/215 -Report_on_the_Telephone_Records_Program.pdf

11 See, e.g., Klayman v. Obama, 957 F.Supp.2d 1, 32–37 (DC Circuit, 2013), https://www.leagle .com/decision/infdco20140116h89

12 See, generally, training log for Business Records (BR) FISA program, declassified and made available online by the Office of the Director of National Intelligence, https://www.odni.gov /files/documents/1118/CLEANED032.%20Basket%202%20-%20NSA%20training.log.pdf

13 NSA BR FISA training log

14 NSA BR FISA training log, slide 8

15 NSA BR FISA training log, slide 8

16 NSA BR FISA training log, slide 11,

17 NSA BR FISA training log, slides 14 and 17–18

18 NSA BR FISA training log, slide 30

19 NSA BR FISA training log, slide 36

20 See, e.g., Elizabeth Mann Levesque, "What Does Civics Education Look Like in America," The Brookings Institution, July 23, 2018, https://www.brookings.edu/blog/brown-center-chalkboard /2018/07/23/what-does-civics-education-look-like-in-america/; Alexander Heffner, "Former Supreme Court Justice Sandra Day O'Connor on the Importance of Civics Education," *Washington Post*, April 10, 2012, https://www.washingtonpost.com/lifestyle/magazine/former-supreme -court-justice-sandra-day-oconnor-on-the-importance-of-civics-education/2012/04/10/ gIQA8aUnCT_story.html?utm_term=.6b11bbac55e3; and "Our Story," iCivics.org, https://www .icivics.org/our-story (accessed April 17, 2020)

21 Katie Miller, "A Lawmaker Wants to End 'Social Media Addiction' by Killing Features That Enable Mindless Scrolling," *Washington Post*, July 30, 2019, https://www.washingtonpost.com /technology/2019/07/30/lawmaker-wants-end-social-media-addiction-by-killing-features-that -enable-mindless-scrolling/?utm_term=.7f6cbb5a2acb

22 Knorr, "The Sneaky Science Behind Your Child's Tech Obsession"

INDEX

ABOUT THE AUTHOR

Photo by Brian Landis

A lawyer by training, **APRIL FALCON DOSS** went to work at the National Security Agency not long after the terrorist attacks of 9/11. She spent thirteen years at NSA, immersed in cybersecurity and the knotty questions that arise at the intersection of technology, privacy, and law. She helped reshape information sharing among US government departments and agencies during the post-9/11 reforms of the US Intelligence Community (IC). She managed counterterrorism information sharing programs with US agencies and with foreign allies. She served as a foreign liaison officer, stationed overseas and managing intelligence oversight for US IC interactions with foreign allies.

Doss was part of the NSA senior management team charged with developing new technology capabilities and led the team that developed NSA's framework for legal, policy, and compliance vetting of the cloud analytics that were designed to draw conclusions from data collected by NSA's signals intelligence programs. She spent half a dozen years in the NSA's Office of General Counsel, where she became the associate general counsel for intelligence law. In that role, Doss managed dozens of attorneys who provided legal advice on NSA's intelligence activities, its new technology development, its filings to the

Foreign Intelligence Surveillance Court, its intelligence oversight and compliance programs, and its privacy and civil liberties programs.

Now working in the private sector, Doss advises clients of all kinds on the complicated data privacy rules that govern their activities. She teaches courses in Information Privacy and Internet Law at a nationally ranked law school and is a regular commentator on issues relating to data privacy, cybersecurity, and national security.